BATTLE TACTICS
TRENCH WARFARE

BATTLE TACTICS
TRENCH WARFARE

Dr. Stephen Bull

COMPENDIUM

© Compendium Publishing, 2003

ISBN 1 902579 72 0

British Library Cataloguing in Publication Data
A CIP catalogue record for this book is available from the British Library

All rights reserved. No part of this publication may be reproduced, stored in a retrieval system or transmitted in any form or by any means: electronic, electrostatic, magnetic tape, mechanical, photocopying, recording or otherwise without permission in writing from the publishers.

Printed and bound in Hong Kong through Printworks Int Ltd.

Previous page: **British troops in the firing line before La Boiselle.**

Right: **German troops wait to be interrogated, Pilckem Ridge, July 31, 1917.**

Below: **The Marine and 1st Naval Brigade of the Royal Navy reinforced ANZAC troops on Galipoli. Here they use periscopes to check on Turkish troop movements. (IWM Q13426)**

Cover photos: **Background picture, see caption page 37. Inset picture see caption page 85. Back cover picture see caption page 37.**

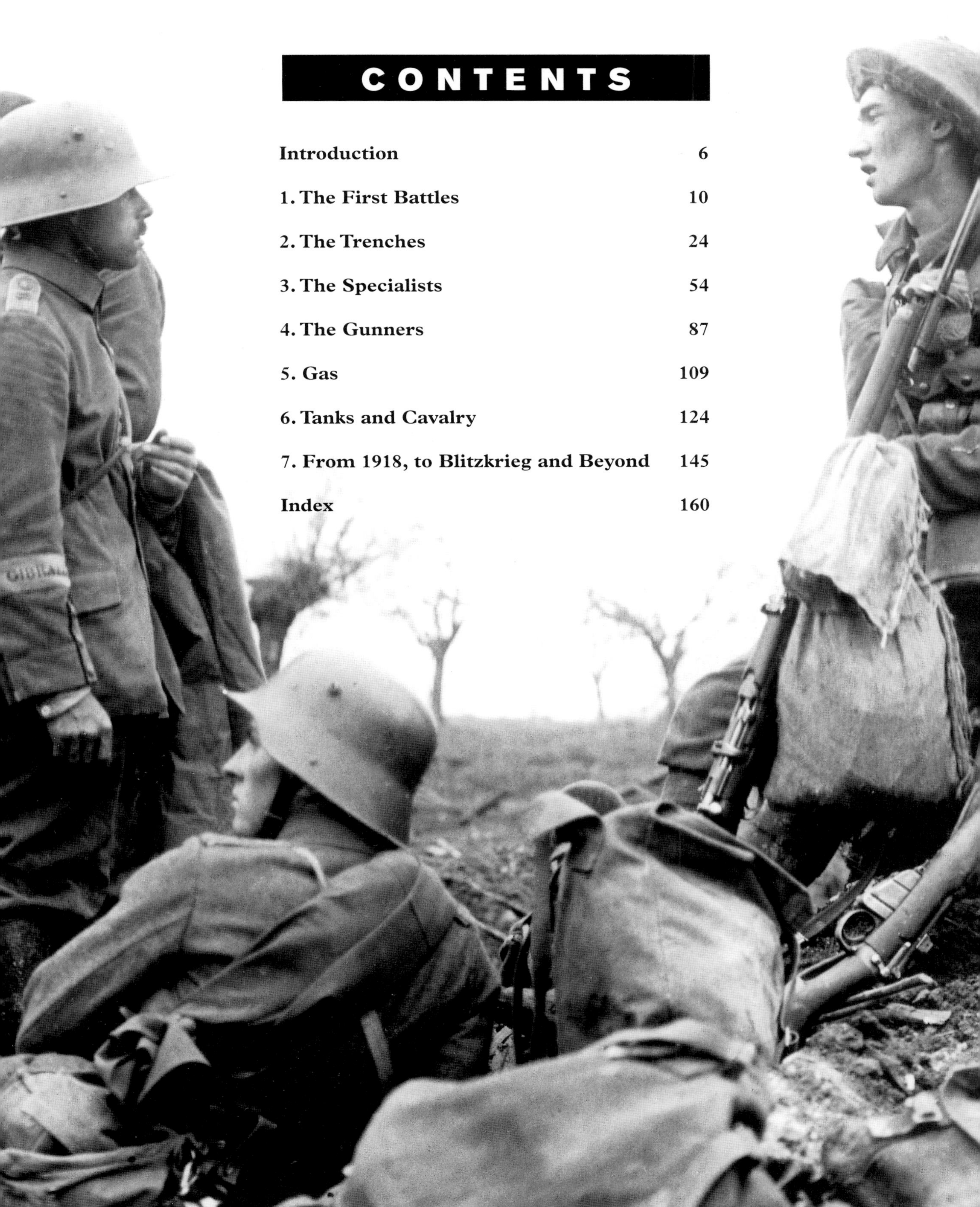

CONTENTS

INTRODUCTION

Earthworks have been a feature of warfare for millennia. From the Iron Age hill fort, to Roman camps, to the defense lines of Saratoga in the American Revolutionary Wars, fighting men have understood the importance of physical barriers and entrenchments. Earth has served to hide the soft human form from missiles, present an obstacle to enemy advance and assault, and give moral courage to the defender.

The importance of the earthwork received an important fillip in the Middle Ages with the advent of gunpowder and cannon. No longer could walls of masonry be expected to protect castles and towns from siege, as was so graphically demonstrated at Harfleur in 1415, and Constantinople in 1453. So it was that the "high thin walls" of the Middle Ages gave way progressively to artillery, the castle eventually becoming the stately home in England, or the equally grand and unwarlike Schloss or château of Germany and France. At first the changes were almost imperceptible, with minor additions to existing stone structures like the medieval adaptations to the walls of towns like Canterbury and Southampton which allowed the mounting of cannon. Later gunpowder would dictate the pace of change, and the layout of defensive plans, as was the case with Brunelleschi's bastions and ramparts in Italy.

By the 17th century stone castles were all but irrelevant to the military scene unless they incorporated earthworks, and these became ever larger and more costly. They would be the major means by which the Dutch would defend their new found independence from Spain, as at Bergen-Op-Zoom in 1588, or Ostend in 1601–04. Even in relatively peaceful England earthworks would play an important part in the Civil Wars. The quality of these defenses could be the deciding factor how long a strategic location could survive, as for example at Liverpool, Lincoln, or Oxford, the Royalist capital, whose 1640's defenses now survive only in painting and plan. Elsewhere archeological investigation has helped demonstrate just how formidable earth could be, as at Newark, or Gloucester, where the foundations of Parliamentarian bastions have survived beneath the city streets. On a smaller scale earth defense would even be exported to the new world of colonial America.

By the age of Vauban in the late 17th and early 18th centuries the bastioned fortress had become perhaps the single most important feature of warfare in Europe. Sieges in which trenches were dug around enemy towns, and up to the defenses in the form of "saps," had become the norm. Soldiers who dug would become known as sappers, and infantry unfortunate enough to take their turn manning trench lines would come to fear the sniper's bullet and the bomb. Soldiers stormed town walls, and from one trench to another with muskets, edged weapons, and the explosive "grenadoes" which had entered the military vocabulary in the 16th century. The best military engineers were those skilled in the arts of traces, gabions, and mortars.

Though fortresses became less dominant in the early 19th century, when cavalry and Napoleon's wars of maneuver took center stage, the art of entrenchment was far from forgotten. Bloody storms marked the end of sieges at key locations. In 1812 at Badajoz in Spain British troops suffered more than 3,000 casualties wresting the town from its French and Spanish defenders after a protracted siege – then went out of control in the ruins. The next year it was the turn of San Sebastian, stormed and bombarded until only the citadel remained to be surrendered by its French defender, General Rey.

If reminder was required of the significance of earthworks, the second half of the 19th century would provide many sanguinary examples. At Sevastopol in the Crimea the siege would last 349 days with the Russian occupants deploying nearly 1,000 guns against an almost equal number of French and British pieces. Moreover it would be the fate of Sevastopol which helped dictate the locations and importance of the field actions of the campaign. Even at Balaklava the dispositions of the troops and the famous, but ill-fated, Charge of the Light Brigade would be influenced by the presence of field works which were captured by the enemy.

In retrospect the American Civil War should have given warning that field fortification and entrenchment

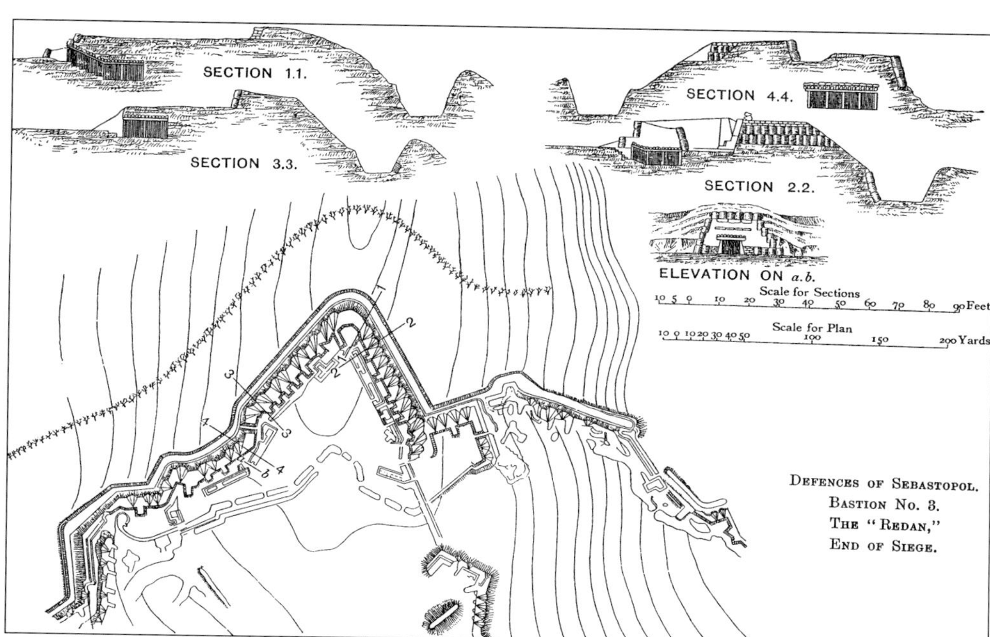

Left: **Trenches, artillery, and obstacles form the main Russian defenses at Sevastopol, 1854.**

were likely to be used on an increasing scale. Even at the beginning West Point was teaching its officer cadets that engineering and the spade were full partners in war with the infantry, cavalry, and artillery. Washington itself boasted defenses 37 miles in length. The Confederate General Lee, now better remembered as a field commander, earned the nickname "Ace of Spades," for the amount of digging he ordered his men to do during 1861. In many instances during the war the combined power of spade and rifle would be put to the test.

During 1862–63 Vicksburg held out against the Union for 213 days, protected by a line of works over seven miles long. These defenses were not just individual forts, but formed a chain linked by rough rifle trenches. Despite bombardment, the Confederates in Vicksburg repulsed repeated assaults, some of which foundered under withering fire even before the attacking troops had reached the perimeter. During Grant's attack in May 1863 the Union infantry had only to cross an open gap of between 80 and 100 yards in width, yet they were decimated by fire. At Petersburg, later dubbed the "Confederate Sevastopol," the works of both sides reached epic proportion. Meade's attacking force lost over 4,000 men in a single assault.

Moreover trenches and rifle scrapes were now finding their way with ever greater frequency into field actions. Where time allowed, holes in the ground were supplemented by logs and cleared fields of fire. It has reasonably been suggested that earthworks were particularly popular in the Civil War not only because rifles and artillery were improving in range and accuracy, but because the existence of strong points had a particularly strong steadying effect upon armies which were made up predominantly of citizen soldiers. Frighteningly informed opinion now suggested that attackers required a local superiority of about five to one to deal with a thoroughly entrenched adversary. As General Jacob D. Cox put it, "one rifle in the trench was worth five in front of it." This deadly combination of new weapons, mass armies, and trenches made this America's bloodiest conflict.

Though the Civil War was widely assumed to be untypical and no guide to future events, the period did offer comparisons in other parts of the globe. In 1864, during the war of the Danish Duchies, the lines of Düppel held out for two months on a two mile front. The Prussians dug parallels systematically closer to the enemy lines, but still lost a thousand men in the final assault. In the Franco-Prussian War, Paris stood a siege for 131 days despite the fact that French field armies were already defeated, and that the main French defensive forts had been designed 30 years earlier. Belfort held out for 103 days, resistance there being terminated only by the cessation of hostilities.

In 1877 the Turkish Army surprised the world at Plevna when it resisted more than three times the number of Russians for five months. At the height of the siege on 11–12 September 95,000 men under Grand Duke Michael attacked Plevna from several directions. The Russians lost over 20,000 without crushing Turkish resistance. Though finally forced to capitulate, the Turks had put up an astonishing performance with the aid of

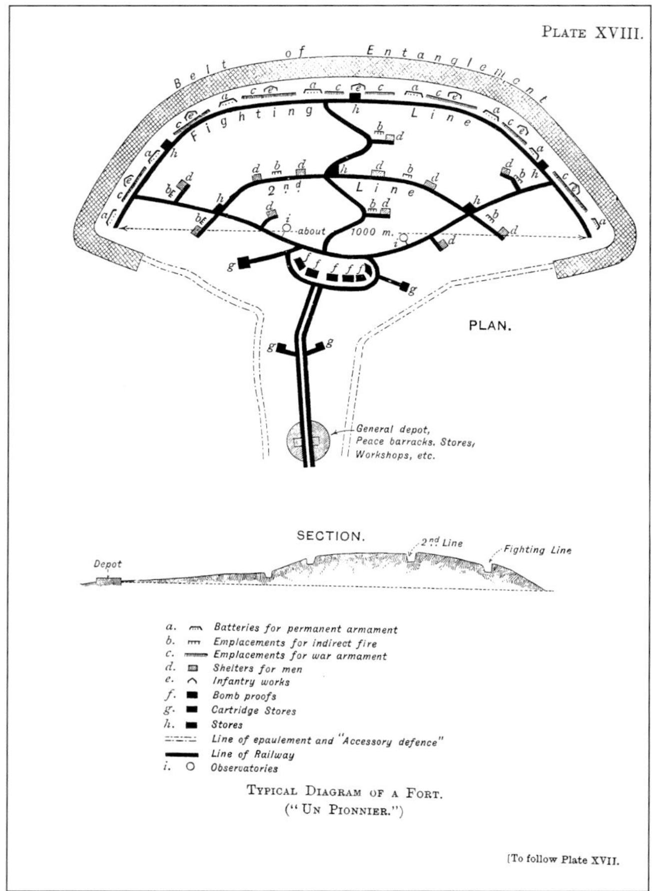

entrenchments. The war correspondent of the London *Daily News* was moved to remark that, "the whole system of attack upon even the simplest trenches will have to be completely changed in future."

Though the basic assumptions about war were not overturned overnight it would be disingenuous to suggest that no lessons were learned as a result of such experiences. By the 1880s French forts were being designed which were essentially mounds of earth protected by artillery, within which were entrenched "fighting" and "second" lines. These were linked by further trenches and fronted by a "belt of entanglement." Within the system were dugouts for shelter, "bomb proofs" and cartridge stores. According to General Brialmont ideal works would now include rifled mortars, armor-plated batteries with the guns supplied by rail lines, and concrete and earth defenses which would take into account the effect of high explosive shells. Even so the French conclusion was that the offensive was superior to the defense, and this would have important consequences in 1914.

The value of such technological defensive preparation appeared amply demonstrated during the Russo-Japanese War when, in 1904, Port Arthur was defended for 154 days. Here the Russians had expended a million roubles a year, over several years, to create 11 main works on a four-mile frontage, connected by trenches and covered by over 200 guns. Repeated and determined Japanese infantry assaults made little headway, and were estimated by German observers to have cost 80,000 killed and wounded. At times the fighting took on the character of full-blown trench warfare with attacks from trench to trench.

How these lessons were assimilated varied from place to place. It was arguably the German Army which took most practical heed. The heavy artillery was augmented, a heavy trench mortar developed, and grenades retained. Britain also took some note of the dispatches of Sir Ian Hamilton and his staff from Manchuria, and this doubtless helped in the selection of artillery equipments. There were also experiments with grenades, and even rifle grenades. Yet it is arguable that the idea of trench warfare did not fully penetrate, and for good reason, since Britain did not expect to fight the sort of war which would require mass armies, nor solid lines of trenches.

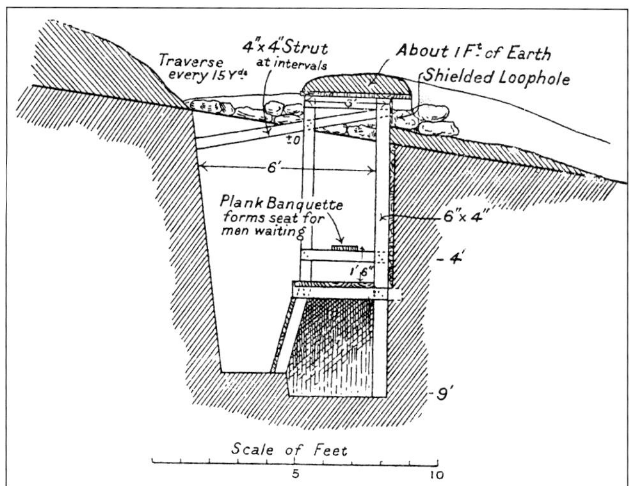

MITRAILLEUSE (BACK VIEW).

Above: **Russian "blinded trench" as used during the defense of Port Arthur.**

Above right: **French mitrailleuse, c.1870. Early machine guns were tactically limited both by their size, and by the fact that they were often deployed as a form of artillery.**

Right: **Boer riflemen in South Africa. The British Army was taught hard lessons on the importance of scouting, skirmishing, and digging in.**

Above left: **French fort design at the end of the 19th century. Considerable similarities are apparent with the continuous trench lines which would appear in late 1914.**

Far left: **Japanese infantry heroics against the Russians. Huge casualties and a surprise Japanese victory sent mixed messages to Western military planners.**

CHAPTER ONE
THE FIRST BATTLES

Kaiser Wilhelm II is said to have referred to the British Army of 1914 as a "contemptible little army," and there was no denying that it was in modern terms "little." At the outbreak of war on August 3d, 1914 the Regular British Army consisted of just under 250,000 men; tiny when one considers the requirements of policing a vast Empire, as well as putting a four division "Expeditionary Force" onto the continent of Europe. Yet, in the words of General Spears, the regular army was, "a perfect thing apart." As a volunteer rather than a conscript, with experience of Empire, the individual British soldier was certainly different to many of his continental counterparts. As one officer of the Northamptonshire Regiment had reflected of the regular in the Boer War:

"Modern warfare is just a bit beyond him, he has neither the intellect of a highly educated man, the instinct of a savage or the self reliance of the colonial. He is a good fellow but a terribly thick headed one. To think for himself is not what he is accustomed to."

However, unthinking obedience could be a virtue in the right place at the right time, and a major premium was put on "steadiness" in the prewar army.

George Ashurst's account of basic training with the Lancashire Fusiliers, at Wellington Barracks, Bury, was in many ways typical:

"On arriving at the barracks we were taken by the sergeant of the guard to a long wooden hut with a stove, forms and wooden beds with straw palliasses. A couple of army blankets were thrown at us and we were told to present ourselves at the barrack dining-halls with the troops at meal times… I soon learned what life was like in the army and that it paid to do as one was told smartly and quickly… I got on very well as a soldier, except for little reminders by the sergeant-major that I was a soldier now and 'Take your hands out of your pockets, stick your chest out and your chin in' as I walked across the barrack square…There was also a school and a teacher in the barracks where one could go in the afternoon and sit at desks with pen and paper to improve one's education… There were quite a lot of ignorant fellows in the army at that time, whereas I was straight out of an office."

If Ashurst had come "straight out of an office" there was good reason to suppose that many of his colleagues had come from the ranks of the unemployed. In good times the army generally got the poorest of physical specimens; in bad times it was ironic that the average quality of recruit went up as fitter and more alert men found it more difficult to gain civilian employment. At times, as during the Boer War, the poor quality of the army recruit had generated official alarm. There was even an idea that the fitness of the nation as a whole was suspect, an area investigated by the 1903 Interdepartmental Committee on Physical Deterioration. Though considerable advances in public health were made in the decade before the war, particularly concerning children, the raw material of the military would remain imperfect. Lord Robert's "Dregs of Society" and Sir Ian Hamilton's "Hungry hobbledehoy" were certainly recognizable caricatures of prewar Britain.

Later statistics, actually gathered on recruits during the war, would show a close correlation between urbanization, industrialization, and poor standards of fitness. While 46% of Welsh recruits, and 44% of recruits from Scotland and the four most northerly counties of England, were passed as Grade I, only 28% from the London area, and 31% from industrial Lancashire could aspire to the same level. A full 10% of men of military age were deemed unfit for any form of service at all.

Though the British Army clung to many old fashioned virtues in 1914, and would not grasp the nettle of conscription until two years later, there were important factors which would help it to weather the storm of the first year. Perhaps the first of these was the Territorial Force system which had come into existence in 1908, whereby most regiments already possessed battalions of part-time soldiers. True these men had to sign on a second time in order to be sent abroad but they were at least partially trained, and ready long before the mass of volunteers who flocked to the colors in the late summer and fall of 1914. It was these volunteers who would form the bulk of Kitchener's Army, the Service Battalions – many of them designated as "Pals" – who would reach the front line in time to swell the army for the great offensive of 1916.

The unfolding of events would be dictated by the Germans and the Schlieffen Plan. This had been established in 1906, the year Count Alfred von Schlieffen retired. It envisioned that seven eighths of Germany's armies would be used against France, while a mere eighth would be used to hold Russia in check. Above all German leaders feared having to fight the fully mobilized forces of France and Russia at the same time, the so-called war on two fronts in which they would be impossibly outnumbered. To win such a war Germany would need to beat first one of these enemies then the other. The reasoning behind attempting to deal with France first was historic and geographic. German calculations suggested that France and Germany, with their well-developed rail systems, could mobilize and deploy their troops in two weeks, while Russia needed at least six. France could therefore be decisively defeated in the first weeks of war and Russia could safely be ignored for this period. (There was nothing to be gained, German strategists believed, from the alternative of a quick attack on Russia. Russia, if sorely pressed, could avoid a serious defeat and retreat eastward as she had done in 1812, playing for time.) This scheme for dealing with Germany's nightmare of encirclement had much to recommend it, yet it was unduly fatalistic in accepting that German diplomacy had failed and that she was surrounded with implacable enemies who were likely to strike, and who had to be dealt with pre-emptively.

Critical to the success of the plan was the attack through Belgium. Initially the German high command attempted to maintain the impression that the war was not against Belgium at all, and that the army was merely "passing through." Yet neutrality had been violated, and this was just the spur that British public opinion would need for commitment to war. Bayonetted babies, raped nuns and burned libraries were not an instrument of policy at this period, but a few instances of atrocity, and the many rumors, were seized on by the popular Allied press.

By mid-August 1914 the British Expeditionary Force (BEF) was concentrating its two corps, each of two divisions, and its cavalry, between Maubeuge and Le

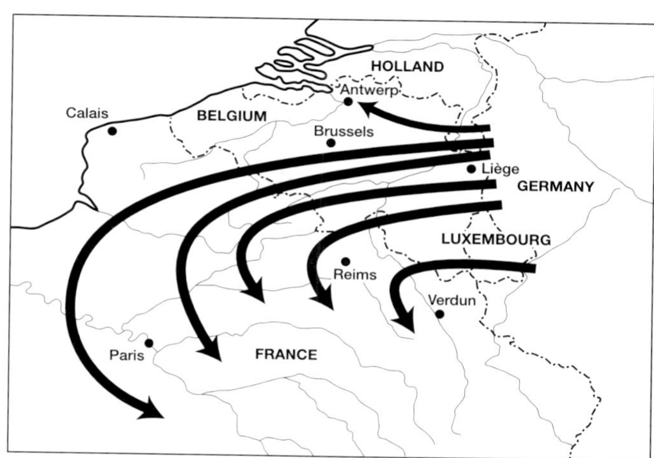

Above: **The Schlieffen Plan: the bulk of the German Army envelops Paris and the French and British armies by swinging through the Low Countries.**

Cateau in northern France. An extra division, numbered as the 4th, followed shortly after. For two vital weeks "brave little Belgium" resisted virtually alone while her would-be saviors prepared. The French launched their ill-fated, and poorly planned, offensive in Alsace. On August 21st the BEF began its advance into Belgium, the intention being that, as part of a general allied advance under the orders of General Joffre, it could interfere with, and possibly even turn, the flank of the mighty German maneuver through Belgium. First contact with the Germans came the next day with the scouting and cavalry skirmishes which were to form the curtain raiser to major battle at Mons. The BEF was standing right in the path of advance of the German First and Second Armies.

The bloody canal defense at Mons on August 23d came as a shock to both sides. Though they were outnumbered, a defensive stance conveyed to the British an advantage which would cost the Germans dearly. The 75th (Bremen) Infantry Regiment alone took 381 losses in a single attack; nor was this exceptional, especially where the Germans presented crowded targets to rifle and machine gun. As the British Official History described the fight at the railroad bridge of Les Herbières:

"The enemy then plied the East Surrey defences with shrapnel and machine gun fire for half an hour, causing no casualties, but disabling one machine gun; after which, about 1.30 pm, he attacked with two battalions of the 52nd in mass, which advanced across the open at a range of 600 yards. Such a target was all the British could wish for:

Left: **Jingoism 1914 style, the British bulldog at Mons.**

Right: **Going to the wars, a Territorial of the Norfolk Regiment in a farewell photograph with wife and child.**

another company of the East Surrey had by this time joined the one astride the embankment, and three platoons of the Suffolk had also come up to cover their left flank. Rapid rifle fire, combined with long bursts at selected objects from the remaining machine gun at the barricade, mowed down large numbers of the enemy and scattered the rest."

Yet the British also suffered, and it was quickly apparent that if they did not retreat they could well be isolated and defeated in detail. Thus began the retreat from Mons, which would lead to a second defensive battle at Le Cateau. Just how swift this retreat was may be illustrated by the fact that while some units, like the 1st Battalion, Queen's Own (Royal West Kent), or 1st Battalion, Northumberland Fusiliers, were very heavily engaged, others hardly saw the enemy. 1st Battalion, South Wales Borderers, for example, of 3rd Brigade, 1st Division, opened serious fire only once, when their D Company spotted a Uhlan patrol. Some units covered as much as 250 miles in the 17 days between August 20th and September 5th; even the luckiest got just two days rest. Private Frank Richards of 2rd Battalion, Royal Welch Fusiliers, with 19th Brigade, would record more hard marching than fighting in his retreat.

"We marched all night again and all the next day, halting a few times to fire at German scouting aeroplanes but not hitting one. At one halt of about twenty minutes we realised that the Germans were still not far away, some field artillery shells bursting a few yards from my platoon, but nobody was damaged. We reservists fetched straight out of civilian life were suffering worst on this non stop march, which would have been exhausting enough if we had not been carrying fifty pounds or so of stuff on our backs."

It has often been remarked that the Germans mistook the rapid British rifle fire for machine guns. Yet this was at least in part a reflection of unrealistic expectations, and a difference in tactical application. The standard British rifle was the Short Magazine Lee-Enfield, or SMLE, a 0.303-inch bolt-action weapon with a ten round magazine. It had been introduced in 1904, in the aftermath of the Boer War, with the intention of replacing the various long infantry rifles and the short carbines of the cavalry and artillery with one "universal" weapon. It was shorter and handier than the equivalent French and German arms, though it was fitted with a suitably fearsome 17-inch sword bayonet for close quarters. Absolutely murderous in the hands of even an indifferent shot at 100 or 200 yards, it was "effective" at 800, and at the start of the war at least, was sighted up to 2,800 yards. The full metal jacket, Mark VII ammunition, would penetrate at least 9 inches of brick, or 18 inches of sand in bags. The Lee bolt mechanism had a relatively short pull, which, though theoretically marginally less accurate than the German Mauser system, was a little faster, and did not require the shooter's aim to be disturbed while he worked the bolt to reload for the next shot. The Mauser was also slightly at a disadvantage with its smaller integral five round magazine, which would require rather more reloading in a protracted firefight.

Against a human target the result of bullets was not always predictable, and varied with range and circumstance. If the victim were lucky, and hit in a place where there was no bone, the "humane" jacketed projectile might pass right through leaving a relatively clean wound. Some would be killed with a small neat hole and little blood. Close ranges and ricochets were a different story. The medical journals recorded exit wounds five inches across, and there were instances of men with shattered heads and spilt brains who took hours to die.

While the technical differences between the Mauser and the Lee-Enfield were, to the layman at least, marginal, the differences in training and application were obvious. For some years, encouraged by N.R. McMahon, one time Chief Instructor of Musketry at the Hythe small arms school, British training had been moving towards an appreciation of volume fire. McMahon, examining historical precedent, had noted that the longbow had proved a more effective weapon than the crossbow, even though the latter was more accurate. It was not enough, therefore, that the troops be trained to hit a target; best results were achieved when this could be done several times in reasonably rapid succession. In the words of the 1914 *Musketry Regulations*:

"Short bursts or rapid fire may surprise the enemy before he can take cover, they favour the observation of results and afford intervals of time for adjustment of sights and fire discipline… a well trained man should be able to fire from 12 to 15 rounds per minute without serious loss of accuracy."

Prewar experiment seemed to validate the argument. In one case a trial of 100 of the best marksmen against 150 average shots showed that when rapid fire was used the larger body of average men had the best of it. Rapid fire also seemed to be the best counter against the rushes of bodies of enemy troops. In another experiment the results suggested that approximately 23% of an attacking unit would be shot down by the time they reached 600 yards away, if the defenders could get off 30 shots in a three-minute period. Looking at the other side of the equation, if the attackers could get within about 150 yards of defenders in a trench, then they, too, could hope to inflict serious damage, even while advancing. Perhaps 18% of the defenders would become casualties as the attackers closed to 50 yards. Such equations were certainly put to dreadful practice around Mons, as was seen first hand by the young Royal Field Artillery trumpeter James Naylor, who observed an officer giving orders under attack:

"He was saying, 'at four hundred… at three-fifty… at three hundred.' The rifles blazed, but still the Germans came on. They were getting nearer and nearer and for the first time I began to feel rather anxious and frightened. They weren't an indeterminate mass any more – you could actually pick out details, see them as individual men, coming on, and coming on. And the officer, still as cool as anything, was saying, 'at

two-fifty... at two hundred... ' and then he said 'ten rounds rapid!' And the chaps opened up – and the Germans just fell down like logs. I've never seen anything like it... "

Although some of the basic rifle training was pedestrian in the extreme, as for example the infamous Naming of Parts, when it came to action the soldier was supposed to make intelligent use of his weapon with a minimum of direction. As the manual put it:

"If all ranks are kept informed of the course of events and led to anticipate occasions for fire action, there should be no need for any words of command other than those which regulate movement, the opening and closing of fire, and the rate of fire, and even these may be dispensed with if the firers are well trained and combine their efforts according to orders issued in anticipation."

Volleys were now only to be used in special circumstances, with bursts of independent fire being the rule rather than the exception, officers only intervening if fire became wild. By 1914 the smallest basic fire unit was the section, of just ten men, under the direction of a corporal. This had recently been reduced from 25 under a sergeant, at the same time as which the number of companies in a battalion had been reduced from eight to four. So new was this change it had yet to be implemented in the Territorials at the outbreak of war. It was significant in that it was at least tacitly admitted that a mere ten riflemen could now produce enough effect to be worthwhile counting as a separate entity. It was also suggested that four companies might be easier to control than eight, and that the new 200-man companies could act as semi-independent operational teams. Yet the infantry section of 1914 was still of very limited tactical importance, and the corporal in charge was called upon to do little decision making. His most significant function was to ensure that his ten men performed in much the same way as the other 15 identical units in the company, which together might make up a line.

If the British approach to volume rifle fire was advanced, the same could not be said about their deployment of the machine gun. McMahon had been one of several advocating an increase in the number of machine guns per battalion prior to 1914, but for reasons of economy, as well as lack of tactical appreciation, this was not to be done until the war was well under way. This was perhaps surprising, since the British Army had been one of the first to take up automatic weapons, and had scored notable successes with Maxim guns on colonial campaigns. The gun's

Above: **The 0.303-inch Short Magazine Lee-Enfield rifle; seen with bolt fully retracted, as it would be when the rifleman chambered a new round between shots. Good design made 15 rounds per minute perfectly possible, without taking the eye off the target.**

Left: **Kitchener volunteers, ready and willing, 1914. The rifles are old model "long" Lee-Enfields, the uniforms are the blue Emergency type issued to volunteers until enough khaki was available.**

inventor, Hiram Stephens Maxim, had even moved to England and set up a factory at Crayford, Kent, where he entered a partnership with the local firm of Nordenfelt.

Though the early Maxim models converted to fire the 0.303-inch cartridge remained in service at the outbreak of war, the standard British machine gun was the updated Vickers. This had been officially adopted in 1912, and in all respects it was at least the equal of any weapon then in foreign service. It was water cooled and fired at a cyclic rate of about 500 rounds per minute, being fed from fabric belts. Private George Coppard described the drill for bringing the gun into action:

"On the blow of a whistle, Number One dashed five yards with the tripod, released the ratchet held front legs so they swung forward, both pointing outwards, and secured them rigidly by tightening the ratchet handles. Sitting down, he removed two metal pins from the head of the tripod, whereupon Number Two placed the gun in position on the tripod. Number One whipped in the pins and then the gun was ready for loading. Number Three dashed forward with an ammunition box containing a canvas belt, pocketed to hold 250 rounds. Number Two inserted the brass tag-end of the belt into the feed block on the right side of the gun. Number One grabbed the tag-end and jerked it through, at the same time pulling back the crank handle twice, which completed the loading operation."

The sights were now flipped up, and once the safety was lifted out of the way the double thumb trigger could be depressed to fire.

Best operation was in short bursts, slowing ammunition expenditure to 200, or fewer, rounds per minute and observing the fire. In a long term "sustained" fire role it was reasonable to expect the Vickers to pour fourth about 10,000 rounds an hour, though after the first few minutes of rapid fire the water in the jacket would boil. Some of the steam would inevitably escape, but most was led off down a tube to be condensed in a can and reused at a suitable opportunity. A fresh barrel might be required about once an hour, but this was a task that a trained crew could complete in two or three minutes. The gun's range was considerable, it being capable of taking effect up to, and even beyond, 2,000 yards. Even in 1914, when it was regarded largely as a supplement to rifle fire, or as a "weapon of opportunity," the machine gun was perceived to have some very useful characteristics. Perhaps first and foremost was the rate and volume of fire that it could put down. Opinion varied as to exactly how many rifles a Vickers was "worth," but commonly a yardstick of about 30 to 40 was accepted. Tests with short bursts against broad paper targets, representing close packed lines of men showed 20% hits at 800 yards, and 43% at 400 yards.

This comparison of rifles against machine guns could not, however, be exact, because there were other important factors to take into consideration. One was the relatively small target offered by the machine gun and its crew; another was the concentration of its fire. While 40 riflemen could form an effective physical screen of skirmishers, the machine gun, if properly laid at an angle to the intended front, could create a beaten zone which was almost impossible for an attacker to cross. Particularly effective was the use of two or more guns whose beaten zones interlocked and overlapped. Experiment suggested that two machine guns firing against a line of advancing troops, each man two paces from the next, 600 to 800 yards away, could inflict about 60% casualties.

The German MG 08 was little inferior to the Vickers, and used with aggression and skill. *Notes From The Front*, a British official pamphlet of late 1914, noted many enemy tactics, some of them distinctly "unsporting." It was observed, for example, that when the German sledge mount or *Schlitten* was carried flat between two soldiers, 'stander copyright.

"Watch me make a fire-bucket of 'is 'elmet."

and the machine gun covered, the whole resembled a stretcher party. British troops might therefore be lulled into not shooting, to their mortal danger.

There were distinctive differences in British and German weapons and their application of fire, but these were such that they tended to balance each other out. British rifles may have been a shade better for the task in hand, with their firers better trained for volume fire, and the Vickers may have been a marginally better weapon than its German counterpart, but this was offset by lack of numbers and imagination in deployment. The British soldier of 1914, though on average likely to be the social and educational inferior of his German counterpart, was almost certainly better with his rifle, and being more likely to be a regular, was often better trained.

It is, perhaps, fortunate that the campaigns of the first two months rarely called upon the BEF to make a large scale attack, for it was in this area that there was most doubt about the army's tactical doctrine, and it must be suspected that, had British soldiers been thrown into action like the French and Germans, they would have suffered a similar fate.

The attack, as outlined in the official manual *Infantry Training* of 1914 postulated a two stage assault, the first

Above: **The skill of the British rifleman as celebrated in a Bruce Bairnsfather cartoon. The line under the photograph says, "Watch me make a fire-bucket of 'is 'elmet."**

Right: ***Durch Kampf Zum Sieg* – "Through battle to victory". The forces of the Kaiser: Army, navy and colonial troops.**

Below: **Rifles of the combatant nations on the Western Front: Top to bottom, the French 8mm Lebel Model 1886; the Belgian 7.65mm Mauser Model 1889; the German 7.9mm Mauser Model 1898. From the *Textbook of Small Arms*.**

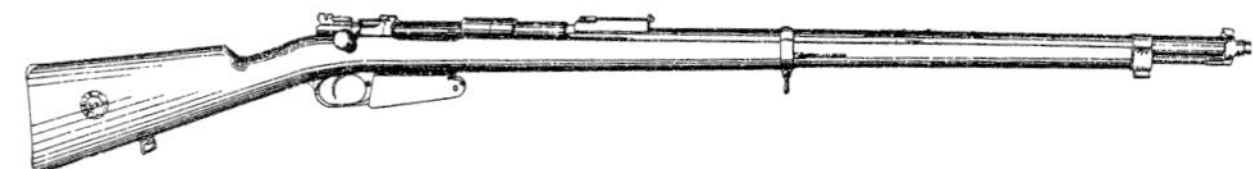

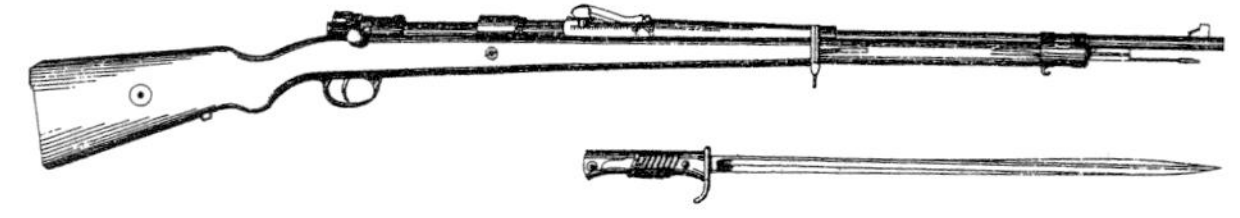

KAMPF ZUM SIEG

of which was to win superiority in the firefight, and the second was actually to close with the enemy. If circumstances allowed, the classic way for a commander to achieve this was to engage with a part of his force only, and then deliver the knockout blow with his "General Reserve." The emphasis was on the final charge with cold steel to rout the enemy, the purpose of artillery, machine guns, and rifles alike being "to bring such a superiority of fire to bear on the enemy as to make the advance to close quarters possible." It was realized, however, that it was unlikely that an enemy would simply be swept from the field by mere impetus:

"Owing to the effect of the enemy's fire, however, this onward movement can rarely be continuous, and when effective ranges are reached there must usually be a fire fight, more or less prolonged according to circumstances, in order to beat down the fire of the defenders. During this fire fight the leading lines will be reinforced; and as the enemy's fire is gradually subdued, further progress will be made by bounds from place to place, the movement gathering renewed force at each pause until the enemy can be assaulted with the bayonet."

Just how this would be interpreted depended not only on circumstance and numbers, but the commanders involved. Brigadier General Ivor Maxse seems to have envisioned that the final push would actually be delivered in a series of waves, on the grounds that a single line could easily run out of impetus. Others had it that the *Infantry Training* manual gave them carte blanche to rush when the opportunity presented itself, and to halt and fire as necessary, a theory which approximates very nearly to "fire and movement." A few went even further, suggesting that only fire would beat down fire, and that more automatic weapons were the obvious answer. At the other end of the spectrum were reactionaries who felt that the whole point of the exercise was that final closing with the enemy, who was not to be beaten by playing at "long bowls."

Although the British Army was called upon to advance on the Marne, it would not be until the Battle

Right: **Sergeant Piper James Stoddard, Liverpool Scottish, and his sons, 1914. Formed at Fraser Street Liverpool, and reaching France in November 1914, the unit was just one of the many Territorial formations which helped maintain the line until the mass of new volunteers were ready for action. (Liverpool Scottish Collection)**

Below right: **The enemy. Men of Infanterie-Regiment Hessen-Homburg, Nr 166, pictured at Bitsch, November 1915.**

Below: **Bayonet practice, an important part of training in 1914.**

of the Aisne in mid-September 1914 that the matter would be put to the most serious test. A German captain observing from the Chivres ridge on the Aisne on 13 September described the appearance of the British attack formations as "a series of waves" of individual "skirmish lines," with ten or more paces between each "funny khaki figure." British accounts noted that battalions were committed on frontages of up to 800 yards, one, or at most two, companies at a time. Against these widely spaced waves the effect of the German artillery was lessened, but against the solid obstruction of trench and machine gun the British lines would grind to a halt.

As the war lengthened and the opposing armies on the Western Front made their "race to the sea" it would become apparent, even to the most obtuse, that this was not a war just like any other. August in this theater saw 14,000 British casualties; September and the fighting on the Aisne 16,000; in October came transfer to Flanders and 30,000 losses. November would add 25,000, while a relatively quiet festive season would list 10,000 killed, wounded, and captured. By comparison Waterloo, almost a hundred years before, but still often taken as the Regular Army's sanguinary yardstick of frightfulness, had cost 6,000. The Old Contemptibles were not actually wiped out, and the numbers killed were usually only a quarter of the total casualties, but wastage and dilution would alter the character of the army out of all recognition.

It was a myth that the old army had died to a man in 1914, and eventually 378,000 1914 Star campaign medals were issued, the majority of them to living recipients. Nevertheless the old army was gone in the minds of the survivors. In the wake of severe casualties new battalions and divisions of Territorials would begin to arrive, and the original regular battalions would themselves metamorphose. Many of the wounded were fit for no further service; others would convalesce and return to a battalion they no longer recognized, or would be kept at home to train others. Officers, if not killed, would be transferred to other battalions to train and lead the inexperienced.

A random example serves to illustrate: In the July 1914 *Army List*, 1st Battalion, The South Wales Borderers, has 26 officers, a handful of whom were loaned to the 3d Battalion (previously militia), while 1st Battalion's own shortfall was topped up with reserve officers. By January 1915 the entry for 1st Battalion is unrecognizable. Though the lieutenant colonel and a major are the same only three or four others are still with the battalion. First killed was Lieutenant M.T. Johnson on the Aisne. More soon followed: Captain Yeatman met a hero's end shortly after he crawled into a wood, armed only with his revolver and dealt with four snipers. Others were wounded, and on September 26th three more officers were killed in a fight around a quarry near Vendresse. Promotions and transfers filled up a total of 11 officer vacancies. At the First Battle of Ypres the Battalion was again in the thick of it and another two officers fell near Langemarck, one of whom was the "very promising young officer" Second Lieutenant Watkins. He had been with the unit ten days. Another officer, Captain Barry, was reported missing but eventually turned up three days later, disabled by a knee wound. More "went west" at Gheluvelt including the veteran Major W.L. Lawrence, DSO, who went down "fighting to the last." Before the unit returned to billets there would be only eight officers left. Back in the trenches in November and December at least one more officer was killed, Second Lieutenant Grylls, who lasted a mere 72 hours.

While enlisted men had a marginally better survival rate, more than 20 men were lost for every officer. In the case of 1st Battalion, The South Wales Borderers, a private soldier still serving with the unit at Christmas 1914 was almost certain to be surrounded by strangers. The battalion had suffered more than 300 killed, wounded, and missing on the Aisne alone, and more than another 200 at First Ypres.

Above: **The war swallowed up much of the prewar regular officer corps. Thousands of "Temporary Gentlemen," like Lieutenant E.R. Bull, 13th Battalion, York and Lancaster Regiment, helped to fill the gap.**

Above right: **British convalescent soldiers pose artistically for a photograph taken in aid of the hospital fund by Fred Whittaker of Stalybridge, 1915. The man with arms folded is wearing a "hospital blue" jacket, the one on the far right a goat skin of the type first issued at the front in the winter of 1914–15.**

Below right: **British Tommies in a fine selection of "liberated" headgear, everything from that of the Prussian Guard cavalry, to a French steel helmet.**

Sickness and detachments accounted for a good number more, so while not all losses were fatalities, the majority of faces in the line were new. Within a week of being sent abroad the battalion received its "first reinforcement" of 10% of its establishment. Other significant new drafts included 150 men received on October 10th, 34 at the end of that month, 196 in mid-November, and 138 on December 4th. Yet the experiences of 1st SWB were the rule rather than the exception. The army was changing, both in character and in tactics, and changing fast.

1929

Left: **Men of 107th Infantry Regiment, US 27th Division, making their way through barbed wire, September 1918 – one man is down already. The American Expeditionary Force first saw action in the form of the 1st Division at Cantigny on May 25, 1918. The first major engagement was the Battle of Belleau Wood, in June 1918. A forest on the River Marne near Chateau-Thierry, Belleau Wood proved the AEF as a serious fighting force – and this view of American strength was reinforced during the Meuse-Argonne Campaign in September–November 1918. But the price was high. Over 300,000 Americans became casualties in World War I, nearly 15% of the AEF's overseas forces. Of these, 53,402 were deaths in battle, 63,114 from causes like disease, and 204,002 were wounded. By the end of the war, 28 AEF divisions had been in battle.** *National Archives*

Inset, Left: **Clad in sweaters and steel helmets, U.S. bombers advance through a gap in the wire. Grenades are carried in bags slung around the neck.** *IWM (Q61343)*

Inset, Right: **A French photograph showing the ruined hunting lodge in Belleau woods. The battle for Belleau involved U.S. Marines and infantry supported by both U.S. field guns and French heavy artillery. The 4th Marine Brigade took particularly heavy losses with the result that after the war the woods were officially renamed the Bois de la Brigade de Marine.**

CHAPTER TWO

THE TRENCHES

The belief that the war could only be won by movement and attack was common to all the major protagonists; yet developments in weaponry had made that movement ever more costly. With the rifles of the armies capable of upwards of ten rounds per minute to effective ranges of over 1,000 yards, and field guns firing ten rounds a minute up to three miles, battlefields had become ever more blasted with steel and lead. If the enemy stood firm, closing with him was likely to prove prohibitively bloody.

Costly wars were not new but what really distinguished 1914, after the stalling of the German offensive on the Marne, was the universality of the deadlock on the Western Front. In earlier conflicts, like the Boer War or the Russo-Japanese War trenches had been dug to besiege a specific locality, or temporarily to defend a relatively short length of a much larger front; now the line would stretch from Switzerland to the sea. This paralysis was not so much the result of strategic bankruptcy on the part of the various high commands, as a function of the effectiveness of modern weapons multiplied by the sheer numbers of men involved, and the relative inflexibility of the rail transport systems on which their deployment ultimately depended. The dreadful equilibrium would be compounded by the near equality of the opposing forces. Moreover space for maneuver had run out. The Leviathans would now only be able to rasp and grind against one another until exhaustion, or technology, or tactical innovation, discovered a chink in the armor.

While it would be 1915 before most generals had admitted that the Great War was qualitatively as well as quantitatively different from what had gone before, "Fieldworks" were accepted as a fact of life. In the British Army while "Siegeworks" were the province of the Royal Engineers and the troops who manned home forts, field works were already a much more widespread phenomena. As the 1911 *Manual of Field Engineering* pointed out:

Right: **Men of the Loyal North Lancashire Regiment trench digging at Tidworth, 1915. (Lancashire County Museums)**

Far right: **Some of the first improvised cover. Belgian troops man a road-side ditch near Nieuport.**

"By field fortification is implied all those measures which may be taken for the defense of positions intended to be only temporarily held. Works of this kind are executed either in the face of the enemy or in anticipation of his immediate approach… field fortification presupposes a defensive attitude and, though recourse to it may under certain circumstances be desirable, IT MUST ALWAYS BE REGARDED AS A MEANS TO AN END, AND NOT AN END IN ITSELF."

It was confidently assumed that such works were a temporary if necessary evil. The main parameters governing construction were terrain, the size and shape of the soldier, and weapon capability. Measurement of the soldier determined that, lying down, he could comfortably fire over cover a foot high; kneeling three feet high; and standing, 4 feet 6 inches. The main opponent of the field work would be the enemy soldier's rifle, which, with the modern *Spitzgeschoss*, or pointed bullet, had the ability to penetrate 15 inches of chalk, 18 inches of sand in bags, 38 inches of hardwood, or 40 inches of earth if it did not contain stones. Early in the war it was not thought that heavy artillery would be deployed specifically to breach field fortifications, and those planning them therefore turned their minds only to the likely effects of shrapnel and splinters from field gun rounds.

Apart from the infantryman's own Implement (Pattern 1908) the tools with which a British battalion was equipped included 226 Shovels G.S., 151 picks, 17 felling axes, eight hand axes, 20 reaping hooks, 32 folding saws, 24 wire cutters, and eight crowbars. The initial provision of sandbags was a mere 30, which perhaps goes some way to explain their scarcity in the first months of the war. Guy Chapman of the Royal Fusiliers noted that at the most dire point of shortage ladies in England sent out to his battalion an extra supply, beautifully hand stitched, and in a range of colors that "the doting Joseph might have dyed for Benjamin," these were obviously too good for the intended purpose and so he kept his boots and wash kit in them. Some "barbarians" in another company did build with theirs but the resulting multi-colored pyramid was conspicuous in the extreme.

Though earth was the main raw material, brushwood, gabions, sacks, sods and timber all had their place in the scheme of things. Barbed wire was a rarity early on, and was thus supplemented with crude

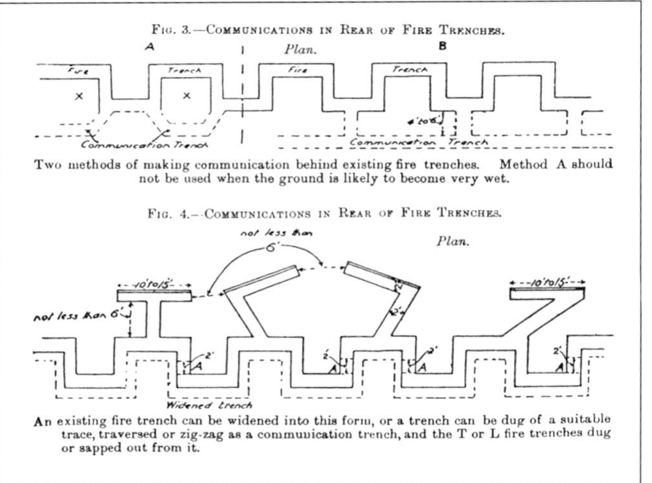

Above: **The ideal trench system, as described in the manual *Notes From the Front*, part III, 1915.**

Right: **How an ideal front line trench should look, with fire step, revetting, neat sandbagging, and an elbow rest. View near Nieuport taken after the Armistice.**

Far right: **Trench reality. Lancashire Fusiliers, one of them wrapped in his ground sheet, engaged in pumping out trenches near Ploegsteert Wood, January 1917. (IWM Q 4653)**

expedients like caltraps, and planks studded with nails. Primitive alarms, like tin cans tied to wires, helped to foster some illusion of security. Actual digging of the field works proceeded on the textbook calculation that, given a shovel and reasonable conditions, the soldier could shift a cubic foot of earth in three minutes. A rudimentary standing fire trench for one could be completed in one hour and 40 minutes, including rests and detail work.

Sensibly the instructions tended to assume that the soldier on the battlefield would spend a fair proportion of his time prone, whether in defense, or coming to the halt at the end of an attack. The first digging effort was therefore "lying down cover," a shallow scrape 6 feet 6 inches long, with the earth piled to the front. Time permitting this would be converted to "kneeling cover" by excavating to a depth of 18 inches, and piling the excavated earth to the front. The most laborious final stage, to create "standing cover" required a further 18 inches of depth. It would therefore be true to say that there was often no clear distinction between the end of open warfare and the beginning of trench warfare, for those engaged in open warfare were often digging, and those engaged in the early weeks of trench warfare would see the exercise as yet another temporary scrape in the ground. Dugouts were few and shallow at this early stage, often being no more than "funk holes" under the parapet. Such efforts would stop very little, but did fulfil the vital criterion of keeping as much of the soldier's body covered as was possible. *Notes from the Front*, published in late 1914, was the first of many booklets produced by the Army Printing and Stationery Service under Captain S.G. Partridge, Royal Engineers, giving official advice. Trenches were ordered to be laid out giving the maximum field of fire, while denying the enemy a clear view of the position. Practical experience had proved that the best form was:

"… deep, narrow, and with low command. The rifle when resting on the parapet, must sweep the ground immediately in front… strong traverses should be provided every four yards or so to localise the effect of high explosive shell falling into the trench, and also give protection against enfilade fire."

Earth from excavations was now placed behind as well as in front of the line. This rear parados both prevented a clear silhouette being shown of the occupants and gave a measure of protection against shells dropping behind. Complete concealment of the trench was difficult but they often ran along hedgerows or lanes so that they were covered from the front. Elsewhere dummy trenches were dug to lessen the fire directed at real positions. With careful cutting and distribution of the earth dummies needed only be as little as six inches deep to fool enemy airmen. A variation on the theme was to clog abandoned trenches in no man's land with wire and branches, turning what appeared to be a welcome shelter into a trap.

These were simple theories, but local conditions and varied application made the original trench line dug in the autumn of 1914 anything but uniform. As Captain J.C. Dunn recorded of 2d Battalion, Royal Welch Fusiliers, near Vicq at the end of August:

"No one seemed to have a very clear idea of how they were to be sited or of what pattern they were to be, so 'A' company dug a length of wretched one-hour shelter trench with our small entrenching tools. Others scraped out rifle-pits in the banks of dykes..."

The next day they were digging again;

"With our entrenching tools, for the few proper picks and shovels in the company tool cart were not nearly enough to go round. We started to dig several times, because Brigade sent order after order amending the previous one, so Major Williams went to brigade to find out what was really

wanted... We had hardly scratched the soil and eaten some bully-beef and biscuit when we moved off again."

Finally the battalion arrived near Jenlain and, borrowing long handled shovels from some French territorials, they dug some "real" trenches, lack of sandbags being compensated by filling the men's packs with earth.

The official *Notes* on trench design and construction were supplemented by private publications including E.J. Solano's 1914 edition of *Field Entrenchments: Spadework for Riflemen*, a lengthy book apparently based on official sources, but well aware of some of the shortages faced at the front. It recommended, for example, that in the absence of sandbags virtually anything could be used *in extremis* including:

"... large stones, bricks, hard coal, steel plates, rails, or joists, bundles of fish plates, railway chairs, bags, sacks or boxes filled with small stones, nails, bolts, chain, earth or sand. Each of these objects, or parcel of objects, should be as heavy as can be moved conveniently, and not less than 30 to 40 pounds weight. Otherwise the cover may be scattered by the impact of bullets and become a source of danger rather than protection to the rifleman."

Solano gave further practical tips. The rifle, for example, was roughly the same length as the amount of earth required to stop a bullet; brick walls were generally bulletproof if more than one brick thick, and made with cement rather than lime mortar; and grinding the edge of the spade created an improvised axe. One of the most important observations was that the trench should suit its surroundings, neat straight edged constructions being avoided in broken ground. *Spadework for Riflemen*, was followed by Edmund Dane's 1915 *Trench Warfare*, a shorter work, but one which widened its audience by means of photographs and an historical perspective calculated to appeal to the non-specialist reader.

Left: **The combined effect of barbed wire and machine weapons. In this instance the dead are Russians in the ice of the Eastern Front.**

By October 1914, with the main German thrust blocked on the Marne, the BEF was transferred north to the left flank of the line so that, together with the Belgian Army, it now occupied the sector roughly from the coast to the Somme. While it was tacitly admitted that this new line would have to be kept for a period of months, and that the strength of the German positions required methods approximating to "regular siege operations" it was not generally believed that this meant the end of open warfare. It was also true that the low-lying terrain of Flanders presented special problems.

Most important was the fact that the water table here was a mere two or three feet below the surface. Excavating too deep, or any significant rainfall, left the trench garrisons standing in a permanent foul soup, which led to trench foot and other illnesses. Drains and pumping achieved no permanent relief. The only solution was not to dig down, but to build up breastworks, forming what were known variously as "parapet," "command," "breastwork" or "box" trenches. These were tolerably secure but required enormous labor to construct. Worse they were painfully obvious islands in the mire which could not help but attract artillery fire. One such section of built up rather than dug down trenches near Armentières in May 1915 was described by Captain F.C. Hitchcock of 2d Battalion, The Leinster Regiment:

"Our trenches appeared to be very formidable; they were duck boarded, and the parapets and paradoses were completely revetted with sandbags. The parapets were six feet high, and the wooden fire steps being one and a half feet in height, gave a fire position of four and a half feet. Owing to the low lying nature of the terrain the trenches were breast works."

In other places previous habitation brought difficulties. Occasionally, where Bantams, that is battalions of men under the previously specified

Below left: **Trench life. Private Edward Billington, Liverpool Scottish, catches up with *Tatler*. (Liverpool Scottish Collection)**

Below right: **South Lancashire Regiment Tommy with improvised trench cooker, 1915. (Queen's Lancashire Regiment)**

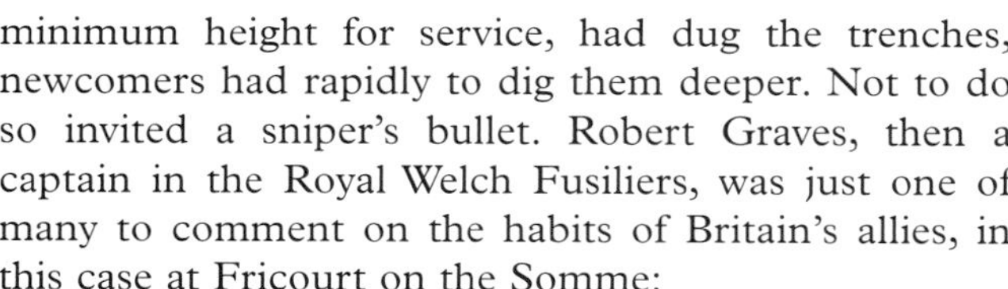

minimum height for service, had dug the trenches, newcomers had rapidly to dig them deeper. Not to do so invited a sniper's bullet. Robert Graves, then a captain in the Royal Welch Fusiliers, was just one of many to comment on the habits of Britain's allies, in this case at Fricourt on the Somme:

"The trenches were wide and tumble down, too shallow in many places, and without sufficient traverses. The French had left relics both of their nonchalance – corpses buried too near the surface; and of their love of security – a number of deep though lousy dug-outs. We busied ourselves raising the front line parapet and building traverses to limit the damage of the trench mortar shells that fell continuously."

The men of Tunnelling Company 174 similarly remarked on the mess, noting that bodies were allowed to lie on the floor of the trench which became progressively shallower, exposing more of the living to the enemy. The condition of French trenches has become a cliché, but there were reasons why they should be so. Most importantly the French were unwilling to admit that a part of their territory was lost. To dig neat, deep, and time consuming works was not only un-French, it was tantamount to an admission of defeat. It was also not surprising that old trenches of any sort should look uninviting; they had been fought over, and now looked cramped and old fashioned compared to the advances in construction being made.

Nor was it only that construction methods improved. New weapons, better transport, supply, and materials all made their mark. Particularly significant from the individual soldier's point of view were the new manning arrangements devised as the war progressed. In 1914 and 1915 long periods actually in contact with the enemy were relatively common. A battalion of the West Yorkshires at Loos spent 70 out of 90 days in the trenches. A Scottish battalion in Flanders recorded 38 consecutive days in the line. Generally this improved with time so that brigades later came to adopt a regime

where only two of their four battalions were actually put in the front and two held back, one in direct "support" another in "reserve." Even within each battalion only half the manpower would actually be in the fire trenches.

Thus it was that, unless a "big show" was in progress, only a small proportion of the army was likely to be right at the sharp end of the war at any moment. When Lieutenant Charles Carrington of the 5th (Service) Battalion, The Royal Warwickshires, analysed his diary for 1916 he discovered that he spent "only" 65 days actually in the front line trenches, with 36 more in support close at hand. A further 120 days were within marching distance, and presumably artillery range. His 101 days in the immediate vicinity of the enemy were divided into 12 "tours" of from one to 13 days.

It is also important to note that the infantry progressively took on many responsibilities, like wiring and revetting, which had previously been the responsibility of Field Engineer companies. From late 1914 the infantry began to form Pioneer battalions, or to convert existing battalions, specifically to take on some of the digging and construction work. By early 1915 the establishment of each division would include a Pioneer battalion, and by the time of the Somme in 1916 labor battalions were also provided by army commands to corps areas. Though often directed by Field Companies of the Royal Engineers these troops did much of the donkey work in building huts, constructing horse standings, and digging dugouts.

Far left: **Lance Corporal with post bag, 4th Battalion, Duke of Wellington's (West Riding) Regiment. Mail was a vital lifeline to normality – for most of the time the only contact with home.**

Above left: **Men of 4th (Territorial) Battalion, South Lancashire Regiment, sit in the mud of a shallow trench, Belgium 1915.**

Left: **Men of the Liverpool Scottish, 10th Battalion, King's Liverpool Regiment, in Q3 trench at St Eloi, 1915. Wooden racks containing Number 2 Mexican grenades and bandoliers adorn the trench walls.(Liverpool Scottish)**

Above: **A French dugout of 1915.**

Above right: **Germans in a bunker or *Stollen* late 1915. The cramped candle-lit conditions are obvious. Kit is stored on improvised shelves, with torches and weapons adorning the beams.**

The Royal Engineers' main specialized tasks behind the lines were now the supervision of new works, rail lines, bridges and water supply, and they also established dumps of supplies for construction and defense. Usually the materials were delivered to the dumps from rail head or warehouse by the Army Service Corps (jokingly referred to as Ally Soper's Cavalry). This work became ever more motorized, the 80 mechanical vehicles which the British War Department had owned in August 1914 becoming 85,138 cars and trucks and 34,711 motorcycles by 1918.

Getting the mountains of material from the dumps into the line was the work of the Brigade Transport Officer, and where carts or light railroads could not go, the motive power was infantrymen, slogging back and forth in "carrying parties." Lieutenant John (later Lord) Reith, of 5th (Territorial) Battalion, The Cameronians (Scottish Rifles), performed the duty of Brigade Transport Officer for 19th Brigade.

"I began to think I was of some importance in the war. I supplied the trenches with a new form of 'concertina' barbed wire entanglement and handled also the first consignment of steel loophole plates… The usual stores included dug-out frames, hurdles, sandbags by the thousand, duck walks and timber of all sizes."

Within the trenches materials multiplied while their issue and stock checking became at once more organized, and bureaucratic. A system was evolved where many items, not routinely handed out to the individual soldier, became "trench stores," to be accounted for by the company and battalion commanders, and left in the trench for relieving units. Gum boots made welcome appearance as trench stores in 1915, steel helmets, later to become a general issue, also appeared in this guise at the end of that year.

Periscopes, both official and private purchase types, were legion but fitted three major categories. The first of these were simple mirrors, with or without such refinements as telescopic arms and mounting clips for bayonets or sticks. Generally more conspicuous but providing infinitely better viewing were box periscopes, in which the mirrors were mounted within a wooden or metal oblong container, sometimes fitted permanently to the trench. Tube periscopes as the name suggests were mirrors mounted within metal tubes, which, while they tended to give limited visibility, were smaller and less obvious. King of periscopes was the Ross, in fairly widespread use towards the end of the war. It was over ten feet long, and fully traversing. Often the only way to make this monster disappear was to mount it inside a dummy tree, or telegraph pole.

Even by the beginning of 1915 the desirability of defense in depth had been realized. Crowding too many men into a single line had drawbacks. They were vulnerable to bombardment, and there was no suitable location to shelter reserves, (medical) dressing stations and latrines. The obvious answer was to build second line or support trenches, connected to the front by winding communication trenches. Very quickly third lines were dug, and in some places defensive belts finally reached anything up to ten lines deep.

In the idealized plan, iterated in publications such as *Notes From the Front*, *Fieldworks For Pioneer Battalions*, and *Notes on Trench Warfare For Infantry Officers*, front line fire trenches were now sturdy and well concealed; they had frequent traverses to minimize both enfilading fire and the effects of high explosive shells. It was not desirable that the plan be too regular because this made it easy to observe, and would take insufficient account of natural features. Furthermore the front line trench should not to be directly parallel to the wire, as this would allow the enemy artillery to calculate its exact location. Instead re-entrants and strongpoints were to be created to allow interlocking zones of fire to surprise the enemy.

The second line was best placed well back from the first so that it could not be overwhelmed simultaneously, and would force the enemy to advance with caution. Even so intermediate positions immediately behind the front were often provided, so that a lost section of the front line could be bombed into, and carrying parties move freely to and fro.

The second line usually contained fewer firing positions and more overhead cover, but it was also the main location of the Vickers machine guns with their posts set at various angles to the front. After 1915 flexible forward defense from shell holes and positions near the wire was the province of the Lewis gun. Second line posts were also established for trench mortars, usually in pairs, rifle grenadiers, and signal stations.

There was no set distance between the British and the German lines. No-man's-land could be narrower than a grenade throw, or further than one could see; yet the distance of the enemy and the degree of danger were not necessarily concomitant as was noted by Private H.R. Smith of the 1st/8th Worcestershire Regiment in July 1915:

"At Hébuterne, so far as danger was concerned, exactly the reverse conditions prevailed to those we had experienced in the 'Plugstreet' sector. There the opposing trenches had been so close they were seldom shelled, as the artillery on the other side were afraid of hitting their own men, but the sniping was deadly. Here, on the other hand, a valley, about seven hundred yards wide, separated friend and foe, and the consequence was that we were frequently shelled, and suffered casualties from this, but there was, at first, little rifle fire. There was a strange, though false air of peacefulness about these trenches in the Valley of the Ancre."

New materials were a major factor in developing trench design. Wooden frames were made up in workshops behind the line for dugouts or revetments, and then moved up whole or in sections for easy erection. Corrugated iron was introduced, with sheets being used for revetting, though it was desirable for small weep holes to be cut so that water would not build up behind. XPM or Expanded Metal Sheets 6 feet 6 by 3 feet were useful, if expensive at £2.50 ($12.50 then) a crate. They could be used directly as revetting, or for the construction of gabions.

Far left: **A cunning alternative to digging down. German troops concealed within the brick stacks at Violanes.**

Left: **Many a true word spoken in jest. Labyrinthian waterlogged trenches were the bane of carrying and relief parties, and could slow movement to a snail's pace especially at night.**

Sandbagging developed into an art form. The most critical factor was that the bags should be filled with the same amount of earth and beaten to approximately the same dimensions – usually three quarters filled and then compacted to a rectangle 20 inches by 10 by 5. Ideally the filling party was three men, two holding and tying the bags while the third shovelled in the earth. Those building with the bags were supposed to work in pairs, and provided that enough hands were available for carrying they were expected to proceed at 60 bags per hour. A significant disadvantage of the sandbag was that it rotted quickly, and it is notable that most modern reconstructions, as at Vimy or the "Trenches of Death" at Dixmude, have opted for the less accurate but more durable medium of concrete. Most British builders used "English bond" in their sandbag constructions – laying alternate courses of headers, at right angles to the face, and stretchers, longest side parallel to the wall. The "chokes" or tied ends of the bag were placed next to the earth leaving, as far as possible, a neat and seamless face on the inside wall of the trench. In places where sandbags were available in quantity the empty bags also became ideal camouflage, not only tied over periscopes and helmets, but concealing loopholes and observation points.

Sods were used in the same way as sandbags and were ideally cut to similar dimensions. Often they were pegged into place using a split picket or other convenient spike. Armor was used in both sniping and observation loops, but contrary to popular belief the plates used were not completely bulletproof as they could be penetrated by the ordinary rifle bullet at point blank range, or by special armor-piercing bullets. They were not normally set proud of the surrounding surface, thereby attracting attention, but into sandbag walls or strong points.

Wiring similarly became a highly skilled business, though the exposure of the wire belts to enemy fire made night work usual, and wiring parties were an unpopular fatigue. Early in the war wooden stakes or pickets were common, but these were progressively replaced with metal types as the war proceeded, some of which could be screwed into the ground. Muffled mallets also helped to avoid detection, but opinion varied as to the best course of action when illuminated by an enemy star shell: One school of thought held that the party should freeze; the other that they should immediately scatter and throw themselves flat.

Neither side was immune to the consequences of detection while on wiring duties, as was related by Private Smith of the Worcesters:

"Sometime about eleven o'clock I could distinctly hear the noise made apparently by the hammering in of metal stakes at a point in the enemy line just opposite where I was stationed on guard with the machine gun beside me. Acting on my orders, I gripped the traversing handles, pressed the double button, and immediately the gun poured its stream of bullets into 'No Man's Land'. I fired about half a belt (125 rounds) before releasing the double button, and did not get a

single jam. When the last echoes of my devil's tattoo had died away I listened for the hammering noise, but it had ceased, and instead there was the sound of shouting from the direction of the German working party – probably someone shouting for stretcher bearers."

Wire obstacles were of two main types: General protective lines, just far enough from the trenches to prevent the enemy throwing grenades in, and tactical obstacles 50 to 100 yards from the trench, designed to channel attacking enemy infantry onto the machine guns. In the best defenses wire was also laid so as to be partly or wholly out of sight, in hollows, ditches, or woods. Very often there were several belts, and sometimes belts were linked by odd strands known as spider wire. Broad areas covered with spider wire were difficult to detect and destroy, and though by no means impassable, they slowed and broke up enemy attacks. The main belts were themselves of several patterns, varying from meticulously planned apron fences to "high wire entanglements" in which pickets and guy wires were lost amidst skeins and concertinas of barbed wire. A few concealed exits allowed patrols in and out of the line. Smaller sections of wire in the form of "knife rests" on wooden frames or "spirals" had special tactical functions in that they could be used to block individual trenches or tracks or even be thrown out of the front line to reinforce particularly vulnerable points.

As the war progressed the bunker and the pill box became more important features of the trenchscape, since defense in depth, and using smaller numbers of men in a series of strong points, proved less vulnerable to the barrage than lining the fire trenches with whole companies. In any case a network of posts was a useful skeleton to any defense – with posts for command, casualty clearance, machine guns, trench mortars and signalling.

The first forms of overhead cover were crude in the extreme. The funk holes burrowed under the front lip of the trench which we have already noted provided rather inadequate cover for one or two men. Officially these rabbit holes met with a mixed reception; they made

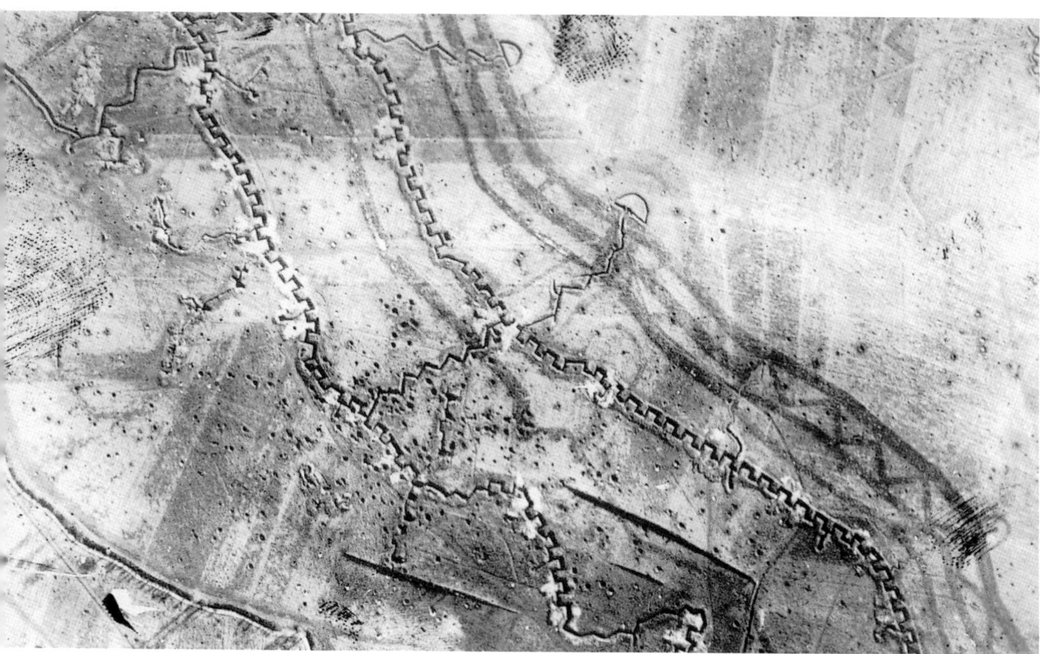

Far left: **The arrival of hot stew at the trenches of the Lancashire Fusiliers, Ploegsteert Wood, March 1917.**

Left: **Aerial view over the front near Cambrai. The first and second line fire trenches are apparent due to their crenelated appearance, while the communication trenches form lazy zig zags. Forward saps cut their way through the shadowy network of barbed wire. The pock marks are shell holes.**

Below: **German stormtroops in action.**

subsidence of the trench sides more likely, and often left legs and arms projecting over walkways to create an obstacle course for carrying parties. Much preferred were the forms of cover described in the *Manual of Field Engineering*, in which square cut recesses between 2 feet and 4 feet 6 inches wide were roofed over with boards, corrugated iron or brushwood and then covered with 9–12 inches of earth. One issue store also used in these constructions were the boxes known as "Cases, Great, Gallery"; these were effectively large crates 6 feet 6 inches high, ideal ready made revetting for an "under the parapet" shelter. Such protection was deemed adequate against splinters, shrapnel balls and grenades. Sometimes protection from the elements was the objective, and more than one sergeant major blanched at the sight of old doors, shutters, and table tops laid across the trenches. Officially speaking four main points were to be observed in the construction of shelters:

"a) The parapet must not be unduly weakened by them;
b) They must not curtail the numbers of rifles available;
c) It must be possible to get out of them quickly; and
d) Simple and numerous shelters are better than a few elaborate ones."

Obviously the degree of protection afforded by such structures was limited, but could make a vital difference as was observed by an officer of the 1st/4th (Territorial) Battalion, The Loyal North Lancashire's:

"It was always a standing wonder that so much metal could fly about in horrid, jagged bits, knocking trenches about, missing men by inches, demolishing dugouts, and yet cause so few casualties. For example, three men were lying in a low dugout with an iron roof; a shell struck the front edge, burying the men and at the same time saving them from its own explosion, which took place simultaneously. Men are sometimes literally struck dumb at these times, as witness the following true story: scene – a slight shelter: officer inside, private at the entrance; three shells fall in quick succession, the first and second miss the shelter by a foot or two and make the usual noise and mess, the third hurtles down and buries itself at the very entrance – a long pause, then a small unnatural voice. 'That's a dud sir!' another pause, another voice of like quality, 'Yes, I see it is.'"

Alternatively fire trenches could be given overhead cover but, unless the roof was propped and loops provided, the defensive function was seriously

Right: **Deep and elaborate German trench complete with sign, *Zur Latrine* – "to the latrine." Where possible stone work was avoided as it became a source of splinters under shell fire.**

Far right: **Fixing scaling ladders to the trench walls before the battle of Arras, April 1917. (IWM Q 6229)**

undermined. Though in vogue in several sectors in all armies in 1915, this system was never common.

Specialized machine-gun posts had been devised before the war. These could take several forms but usually had in common a shallow platform on which the machine gun on a tripod was placed. Behind the platform was often deeper cover giving protection to the crew, or even a trench leading to a nearby splinter-proof shelter. Where the MG position was located in a trench the outline of the pit was often the arc of a circle, which gave a wide field of fire. In all cases they were best placed where they could bring powerful enfilade fire, creating beaten zones through which attacking troops would have to pass.

It was soon realized that it was usually best to keep the Vickers guns out of the front line fire trench, and by 1915 they were normally kept back in or near the second line, where each gun would be given several firing points. The best MG posts were equipped with two or more small bunkers for belt-filling and crew quarters. During bombardments the gun could be dismounted and taken below. Concealment of the MG post was almost as important as protection, and enfilading positions were especially useful in this respect because they were difficult to observe from the front. In well prepared positions the loophole was sometimes concealed by means of a wooden box, the hinged outer face of which was covered by a dummy sandbag. According to *Notes on the Tactical Employment of Machine Guns and Lewis Guns* the MG loop needed to be not less than nine inches in height and should avoid straight edges wherever possible. Where appropriate, grass, straw or gauze blending into the surroundings could be used. Sometimes, particularly during advances or retreats, it was necessary to deploy MGs in open pits; in this case the minimum dimension for the satisfactory operation of the gun was an emplacement four feet square. Specialised posts were similarly constructed for trench mortars, rifle-grenade batteries and signal rocket stations, but in all these cases the fire was indirect, so no loop was required. From the front such positions were either made to be as nearly invisible as possible, or identical to nearby sections of trench.

Sniping and observation points were as many and various as ingenuity allowed, although many were saps dug forward of the main line, sometimes with T-shaped ends. Snipers and observers also made use of ruses. Popular among these were concealed loopholes, as for example where a hole was cut through the parapet and obscured with trench debris, with the rifle muzzle covered by an old tin can or shoe. In another version a single brick would be removed from a wall and the hole concealed with painted gauze. Usually a dark cloth shrouded the back of a loop to avoid throwing the user into silhouette, but where this was not provided, grass, rags or a haversack were kept in the hole. Steel loophole plates made their appearance as early as 1914, and would later become issue items.

Like listening posts or forward shell hole defenses, sniping posts were often reached by a sap or tunnel out to the front of the main line. As Herbert McBride, a sniper with 21st Battalion, Canadian Expeditionary Force, recalled, these posts were usually set at an angle to the front to avoid detection. The clever sniper used several locations to keep the enemy guessing as to his whereabouts.

As the war progressed heavier guns were used in ever greater bombardments; over half of all casualties would come from artillery. Often sandbagged tin and timber structures were not equal to the task. One of many similar incidents was witnessed by Lieutenant Edwin Campion Vaughn of the 8th (Service) Battalion, Royal Warwickshire Regiment, near Herbecourt, in early 1917, when an unfinished shaft of an NCOs' dugout received a direct hit:

"The last of the shells had obviously burst inside the shaft, for the entrance was completely blocked, and the top of the shelter was lying across the trench. In the feint light that still remained we saw the sandbags and pieces of timber half buried in the mud. Holmes stooped to raise one of these short beams, then let it go, with a shuddering exclamation, for he had bent back an arm with Sergeant's stripes."

Four NCOs had been killed by a single shell, one blown to pieces and another impaled on his own rifle.

The obvious protection was to go deeper into the ground, but to do so carried special dangers from gas, lack of ventilation, and difficulty in reaching the surface during attack. Cellars, mine workings and brick stacks had all been used in an improvised manner since the

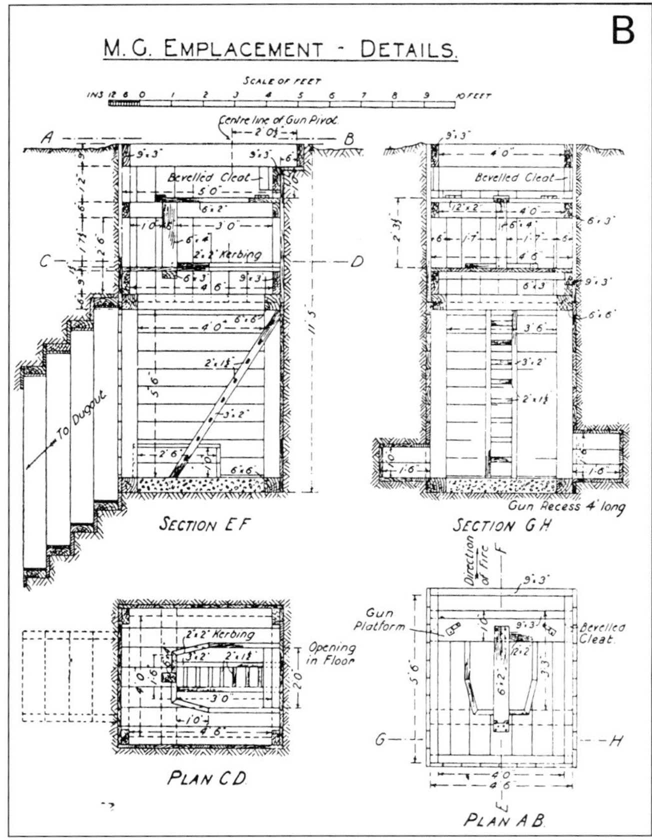

Above: **Machine-gun emplacement attached to deep cover dugout, as depicted in *Fieldworks for Pioneer Battalions*, 1918.**

Right: **How to construct a "bomb store" in the side of a communication trench.**

earliest days, but by the second year of war new plans were beginning to revolutionize dugout design. Curved corrugated iron sheets known as "elephants" were in use by 1916, and had a number of uses. The "large elephant" of 1918 was an arch 6 feet 5 inches high and 2 feet 9 inches deep. When 21 such sections were bolted together with overlaps the "Large English Elephant Shelter" was created, 9 feet wide and 17 feet 9 inches long, a fraction bigger than the earlier "large" shelters. The "Small Elephant" was similar in construction but only 3 feet 9 inches high, 5 feet wide, and 12 feet 9 inches long. Normally these were dug into the earth to form the roof of a chamber. The Royal Artillery specified that its battery officers' dugouts be made from the Small Elephant, strengthened if need be with a frame, and covered with sandbags of stone or broken brick topped off with earth.

Dugout frames were increasingly prefabricated by engineers, pioneers or local factories, and moved up the

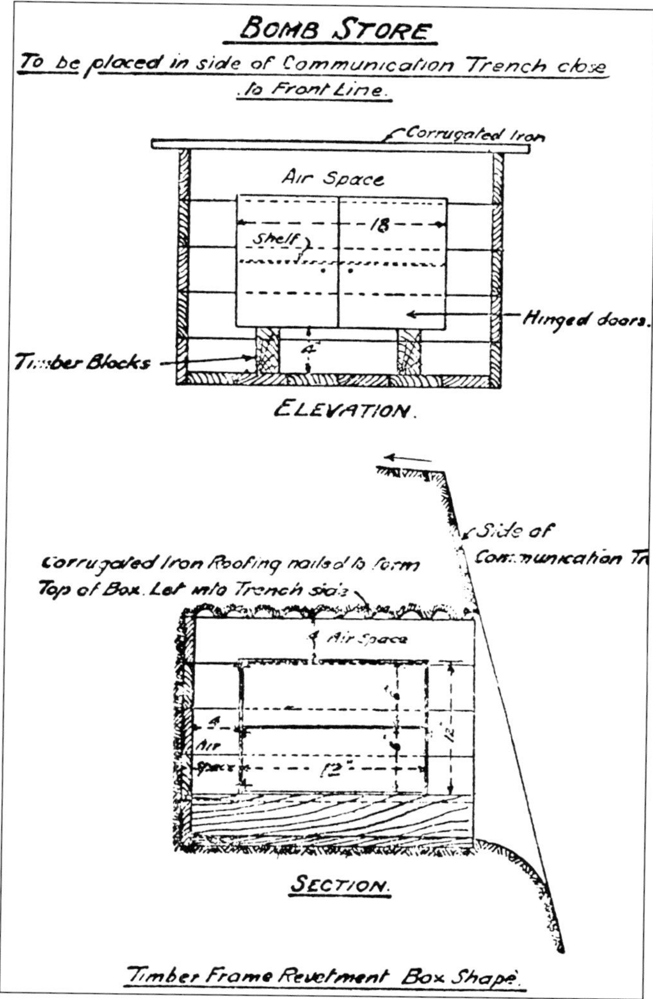

line in kit form. The best were wood lined, equipped with bedsteads, and telephones – sometimes officers even installed phonographs for entertainment. Even so damp remained a perennial problem, particularly in the deeper shelters. There were also accidents with suffocation, as chimneys to ventilate cookhouses, and the "fug" of braziers in sleeping shelters, were problematic. Vents not only revealed positions, but degraded the integrity of overhead cover, and were an invitation to grenades.

Sheer depth was not the only counter to high explosive shells, nor sometimes the best. Heavier shells could penetrate the ground several feet before exploding to form a deep crater. Among the most feared was the 150mm field howitzer shell, known to the British as the 5.9. What evolved as the textbook counter was a sandwich construction with different layers to burst the shell, catch splinters and absorb shock. The ideal "5.9 shelter" had five layers:

1) A bursting course to explode the shell, consisting of two feet of chalk or rubble etc.
2) A cushioned shock-absorbing layer of three feet of soil.
3) A distributing course to spread the shock, of logs or rails.
4) A second cushion layer.
5) A thin splinter proof layer of brick, concrete or corrugated iron.

More than one entrance was best, provided if possible with long approaches and dog legs. Given the right conditions, protection against all but super-heavy shells was, therefore, possible at a depth of about ten feet.

The trenchscape was like a new country, with a new lexis or language all of its own. Rats, lice, and mud were only a fragment of the picture. Many soldiers were alienated when they went home, feeling that even the most sympathetic civilian did not understand. Yet this was not simply a matter of failing to comprehend horrors and bloodshed; there was no way a surface dwelling civilian could understand the new way of life which was unfolding month by month, parts of the mysteries of which were still secret from the fighting soldiers themselves.

It was difficult enough for junior officers in charge of a section of trench to remember what it was they were supposed to be doing. One of the many attempts to remind them was issued as Trench Standing Orders, by 124th Infantry Brigade, in the middle of the war. Within 24 hours of arrival a company commander was supposed to submit a report on the following matters:

"Garrison of the trench.
Field of Fire.
Distance from the enemy's trench.
General condition of the trench.
Whether every man has a post from which he can fire at the bottom of our own wire entanglements.
Number of efficient loopholes.
Whether the parapet is bullet proof throughout.
Whether sufficient traverses.
State of our wire.
State of enemy's wire.
Drainage.
Number of boxes of reserve ammunition.
Number of bombing posts and bombs with each.
Number of rounds of Very pistol ammunition.
Number of Vermorel sprayers.
Number of gongs."

There were also lists of other "trench stores" to be checked, and the officer had to keep abreast of which

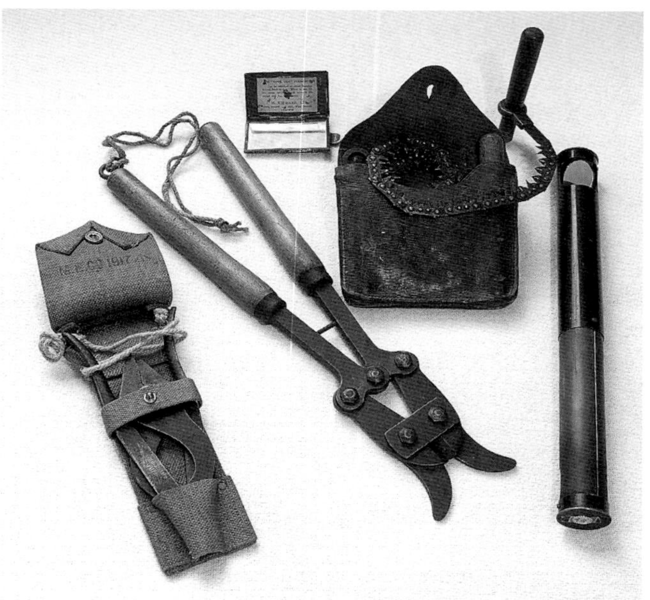

form of store was personal to the man, which belonged in the trench, and which were to be handed over to reliefs or kept with the unit. Sentries had to be posted, at the commander's discretion, but brigade recommended one per three bays by day, one per bay at night or during fog or snow when danger of attack or raid increased. Organising cooking, gas precautions, officers of the watch, rubbish and latrine bucket disposal, standing to arms, rifle inspection and company meetings swallowed another large portion of the company commander's time, but not as much as the "daily return." This included such minutiae as having a record of the wind speed at 5.15 a.m. and 4 p.m., lists of strengths and casualties, and a note of enemy aircraft movements. Discipline was also a headache, even if the men were sober, responsible, and willing to please, as 16 extra rules were in place during active duty in the trenches for 124th Brigade. These included injunctions against sentries who wore anything over their ears or did not remain standing, washing anywhere but in the support trenches, and even hanging up bandoliers in the trench.

Life in the trenches was uncomfortable most of the time, but it is indisputable that the very cheek by jowl existence of the men helped to form a unique sense of comradeship, which excluded many civilians. This is perhaps most vividly captured by some of the novels of the war, like *Im Westen Nichts Neues*, known to English speaking readers as *All Quiet on the Western Front*, but it was also voiced by descriptions penned at the time. This sense being "in it together," was certainly felt by one soldier of the 22d (Service) Battalion, Royal Fusiliers, in the dugouts near Givenchy:

"This sand bag abode is feebly illuminated by a candle dimly burning. My neighbour, who is yet more uncomfortably cramped up, is falling off to sleep, and his muddy, unshaven and jam smeared face is resting on my shoulder. Occasionally he grunts vigorously, making the paper I am writing upon flutter. I've just removed an open tin of jam from under the mud clotted boot of the fellow opposite me. A fair sized piece of cheese is pinned to the sandbagged wall by means of a cartridge. The bread has all been devoured, but a few broken pieces of hard tack biscuits lie scattered somewhere on the ground beneath this living, semi-sleeping entanglement of men. A bayonet thrust in the wall serves as a candlestick, and the candle grease is slowly but persistently dripping on the fellow's forehead who is sleeping directly beneath... And so here we are, huddled and interlaced together, strangers all, until we met in this common cause: in this circle of six we once represented such a different

Above left: **Tools of the trenches, two sizes of wire cutter, a folding saw, and two compact models of trench periscope.**

Above: **Short Magazine Lee-Enfield rifles, seen with an ammunition bandolier, and with breech cover against the mud of the trenches. The first breech covers were old socks, later they were made locally of French cloth, finally they were manufactured in Britain.**

Above right: **Notes from *Instructions on Wiring*, 1918, showing knife rests and low wire entanglements.**

Right: **The layout of obstacle zones, and the dreaded high wire entanglement – a forest of barbed wire 20 feet wide.**

50 YD. STANDARD LOW WIRE ENTANGLEMENT.
Enemy.
52½ yards.
2 yds
2 yds
3 yds
Plan.
Inclined Wire.
In plan
Top end
Low end
Section. (Enlarged.)
Order of putting on the wire.
2 yds
2 yds.
Enemy
LARGE KNIFE REST
4'0"
18"
10' to 15'

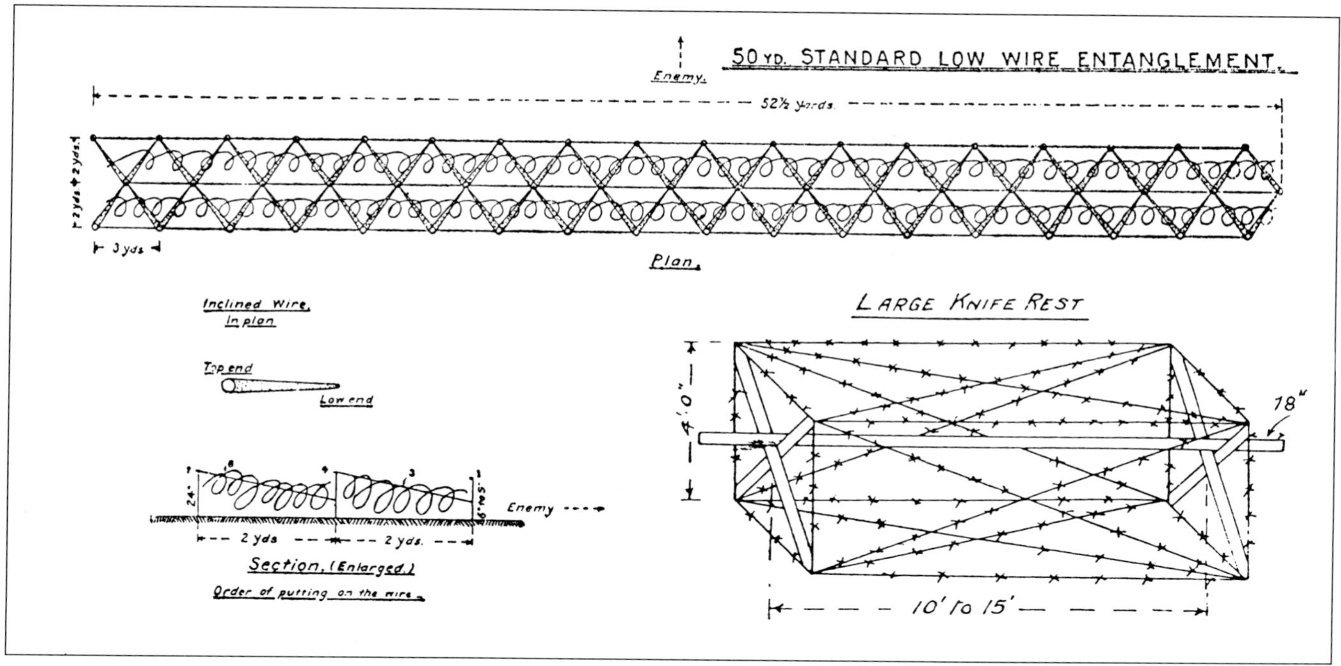

HIGH WIRE ENTANGLEMENT.
8'-0"
Top of Pickets to be waist high i.e: 3'-9" minimum
3" diam: 6'-0" Pickets
Guy wires taut, all other wire to be moderately slack.
3" diam 2'-0" Pickets.
One single coil loose wire, special care to be taken not to waste wire.
6'-0"
2'-0"
1'-6"
SECTION.
PLAN.
METHOD FOR FIXING HORIZONTAL WIRE ON WOOD POSTS
windlass here
running end
1st SKETCH SHOWING BIGHT.
2nd OPERATION SHOWING POSITION READY FOR WINDLASSING WITH IRON BAR.
3rd OPERATION COMPLETED.
AFTER WINDLASSING.
TYPICAL OBSTACLE ZONE
Infantry gap.
Infantry gap
Small copse well wired
Road
Knife rests
Approx: Scale
yds 100
50
0
100 yds
High wire entanglement
Spider wire

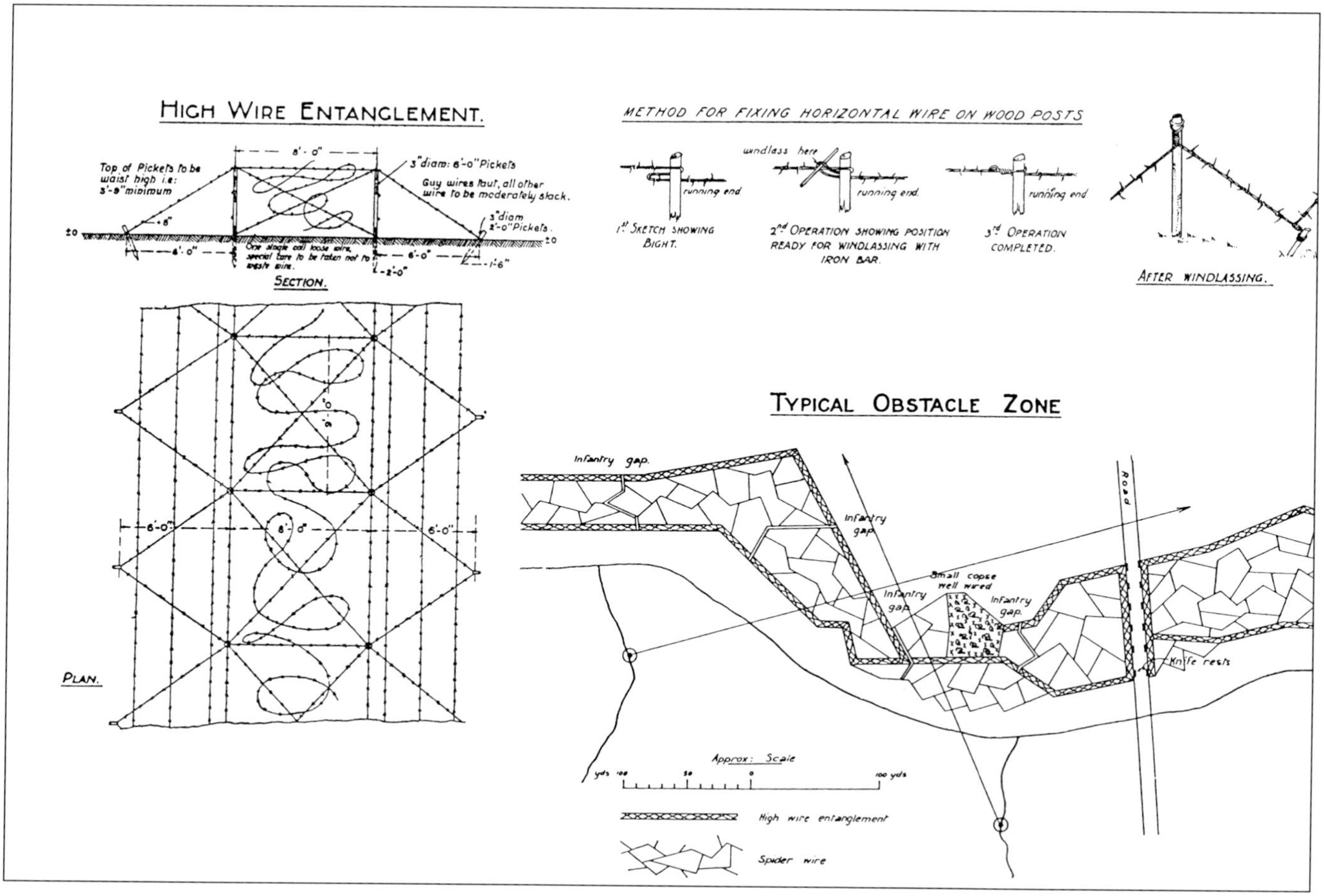

type, but now all gradually approaching the same – the man in his primitive stage."

Coming into, or leaving, the trenches was an elaborate ritual in itself. Platoon commanders were supposed to visit their new homes at least a day before the relief, meet the departing officers and glean useful information. Next the snipers, bombers, and signallers would proceed to their new places. Once this was done the new battalion could march up to the line, and change places with the men in the fire bays the next day.

Attempting to explain what the war in the trenches was all about to the population at home was seldom attempted in any detail, but there were honorable exceptions to the general rule. Books like Edmund Dane's *Trench Warfare* have already been noted, but perhaps the boldest experiments were the reconstructions made in various parts of the United Kingdom. At the seaside holiday resort of Blackpool were the "Loos Trenches." These, billed as "a perfect replica" of the real trenches at Loos in "practically every detail," were both an attraction and a method of raising funds for the King's Lancashire Military Convalescent Hospital at Squires Gate, South Shore. The trench system, which adjoined Watson's Lane, was dug by troops as training before their departure for the front, and maintained by the staff and patients of the hospital, who also provided guides for the visitors. The penny guidebook explained, with an air of awe and showmanship, the appearance of the "empty battlefield" and the purpose of the trenchscape,

"*One of the first impressions to be formed by the visitor on entering the field from Lytham Road, will be the apparently ordinary aspect that presents itself to the gaze. In no direction is there anything to suggest in the slightest manner the all-important fact that almost at their very feet are a couple of miles of admirably planned and cleverly constructed trenches, a perfect reproduction of this astonishing aid to modern warfare. Their wonderful utility in the present war has been frequently emphasised and highly praised by our gallant officers and men; and apart from anything else, they have opened up a new sphere in military operations.*"

Perhaps less successful because of their urban environment, lack of connection with the wounded soldiery, and limited scope, were the representations of

the front constructed in London. In Trafalgar Square was put up a ruined village; in Kensington Gardens was dug a "model trench." These were unrealistic enough to be a source of amusement to soldiers on leave.

As trenches filled the battlefields of France and Flanders from horizon to horizon there were always those who were seeking new ways through them, and while bombardment and frontal attack would continue above ground, there was also a war underground. The idea of "Moles" who would tunnel under the enemy lines was actually raised with the War Office as early as December 1914 by Norton Griffiths, a mining and drainage engineer who had served not only in the Royal Horse Guards, but the Australian 2d King Edward's Horse. Yet the real spur to act came from the Germans, who exploded "mines" under the British lines at Festubert, as a preliminary to conventional attack, later the same month. In retaliation the British Army set the Royal Engineers' Field Companies to mine digging; Norton Griffiths and many like him were quickly recruited, with special appeals directed at the mining communities of South Wales, Durham, and Yorkshire. Some of the best military miners were the so-called clay kickers, men who in civilian life had dug tunnels for sewers and underground railroads. The name came from their technique of lying on their backs, on boards in the tiny galleries, and propelling their sharp pointed spades into the face of the clay. Soon all this manpower would be reorganized into specialized Tunnelling Companies.

Their first major effort came at Hill 60 near Ypres in March and April 1915, when several timber-lined shafts and galleries were pushed under the enemy line through waterlogged ground, using somewhat antiquated tools and methods, by Tunnelling Company 171. Despite enemy counter-mining Hill 60 was torn apart by the explosions, but a German attack would soon force the Royal West Kent Regiment from its summit. From then on the British tunnellers would strive to become more and more proficient, but it is noticeable that the printed material which headquarters provided for their instruction was largely translations of, or observations on, the German efforts. Such was the case with the publications *German Land Mines*, of 1915; *German Mining Officer's Diary*, of 1916; and *Instructions For Mine Warfare*, of 1917.

While professionalism and equipment steadily improved, nothing could really prepare even

Far left: **Men of the York and Lancaster Regiment taking up wire for a night working party Oppy-Gavrelle sector, 1918.**

Left: **German troops in the attack.**

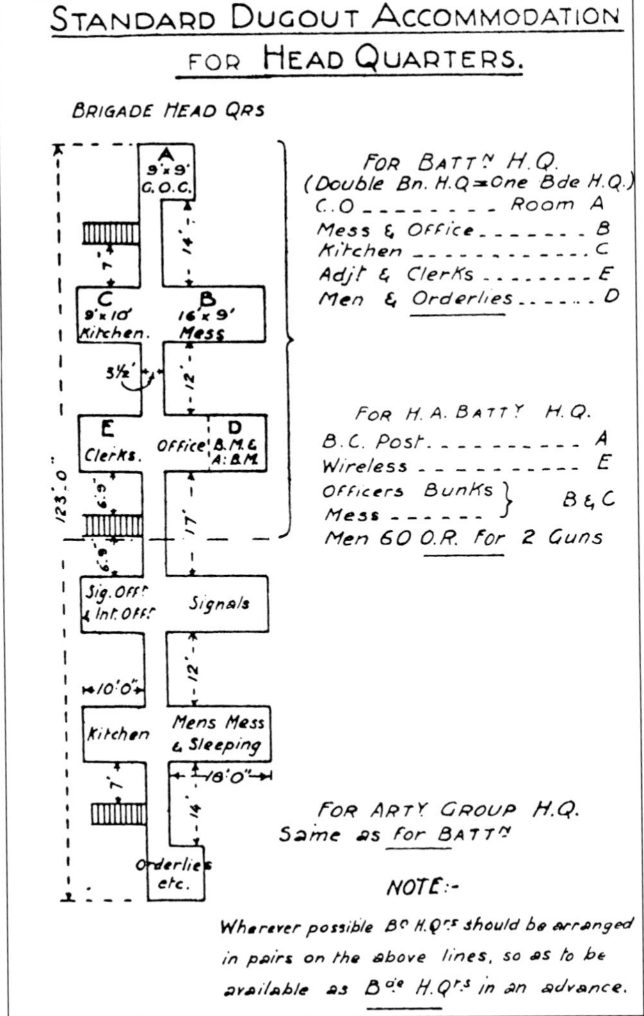

Left: **Standard command dugout plan from *Fieldworks for Pioneer Battalions*, 1918. As can be seen, it was intended that the same basic building blocks of rooms could be used to make up different types of HQ.**

Right: **German diagram showing the war underground. Tunnellers dig towards the enemy line to place large mines, while defenders attempt to frustrate them with smaller detonations which may kill or trap them.**

experienced coalminers for the terror of warfare underground, and the ever present possibility of the detonation of enemy mines designed to collapse the tunnel in which they were working. This was especially the case during "listening checks," when even the pumping of fresh air to the workers ceased, as was described by Lieutenant Geoffrey Cassels of Tunnelling Company 175, near Hooge in 1915:

"Forehead pressed to the face, side or floor of the gallery, one stood, knelt or lay – listening, listening, listening. Sometimes sounds would be heard, dull and muffled. There was always that fraction of a second doubt – when it might be the enemy mining. One's pulse rate would quicken and fright push to the fore in one's whole being… "

There were even instances where the tunnels of friend and foe actually met underground, and fierce struggles would ensue with revolvers and mining tools. Listening would later be improved with "geophones," or stethoscope-like devices, but the work remained as hazardous as ever. The view of other troops on the whole proceedings was ambivalent; tunnellers were a protection in that they intercepted German mines, but they also attracted attention, absorbed manpower, and created huge craters which might have to be defended or assaulted.

One of the tunnellers' most spectacular efforts was timed to coincide with the opening of the Somme offensive. Companies 174, 178, 183, and 252 were all involved, and between them they dug seven large and many smaller burrows under the lines. Among the monster mines were Hawthorn, La Boiselle, and Y Sap. This last had two charges, one of 36,000lb, the other of 24,000lb of ammonal.

Hawthorn was blown at 7.20 a.m. on July 1st; VIII Corps commander Lieutenant General Sir A.G. Hunter-Weston had reasoned, possibly dubiously, that this would give the assaulting Lancashire Fusiliers a head start. The whole performance was recorded on movie camera, by Lieutenant Geoffrey Malins who, to make sure he captured the vital moment, had begun winding his hand cranked camera before the actual explosion:

"I looked at my exposure dial. I had used over a thousand feet. The horrible thought flashed through my mind, that my film might run out before the mine blew. Would it go up before I had time to reload? The thought brought beads of perspiration to my forehead. The agony was awful; indescribable. My hand began to shake. Another 250 feet exposed. I had to keep on.

Then it happened. The Ground where I stood gave a mighty convulsion. It rocked and swayed. I gripped hold of my tripod to steady myself. Then, for all the world like a gigantic sponge, the earth rose in the air to the height of hundreds of feet. Higher and higher it rose, and with a horrible, grinding roar the earth fell back upon itself, leaving in its place a mountain of smoke."

A similarly huge undertaking was the undermining of the Messines Ridge on June 7th, 1917, preparation for

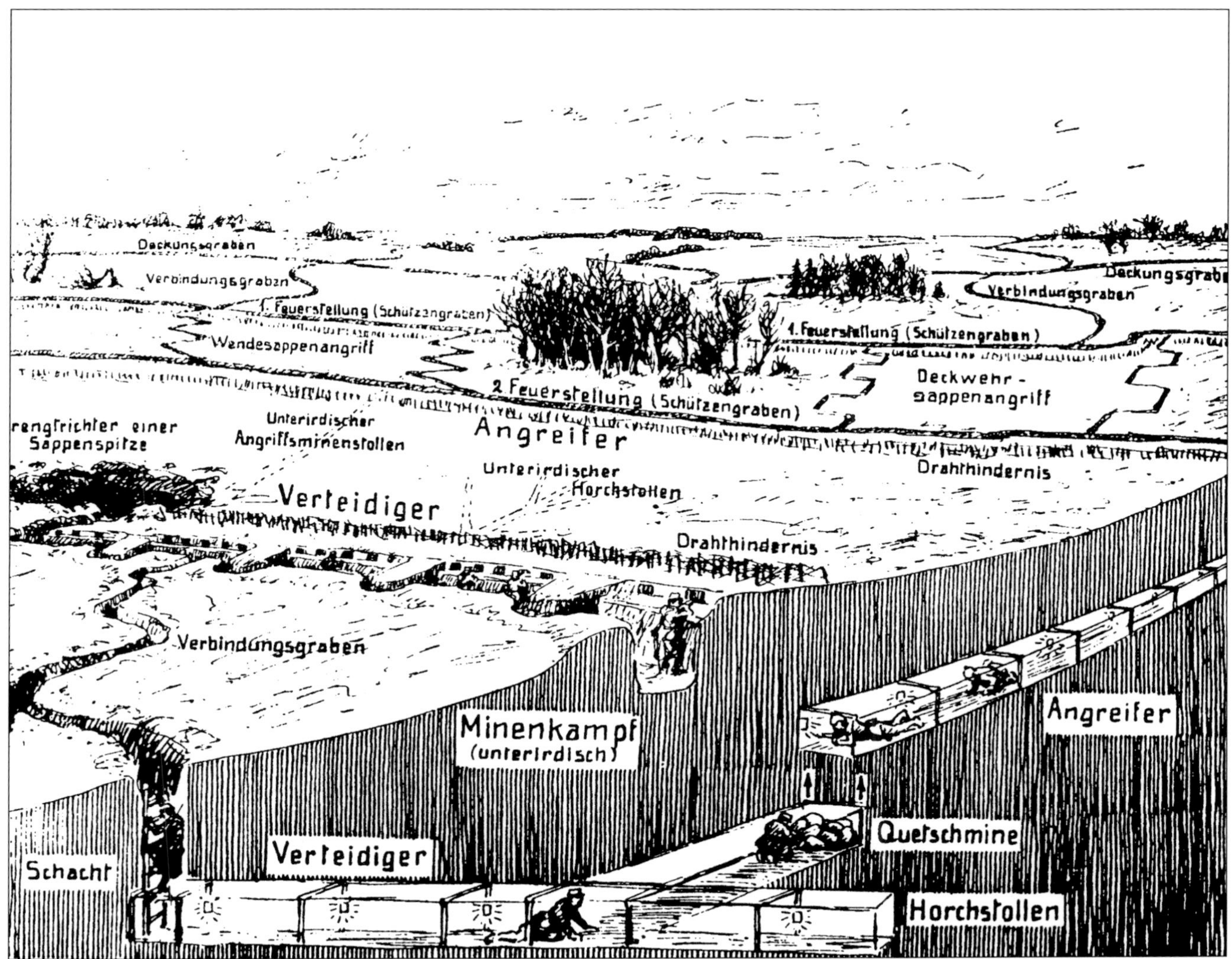

which was started many months earlier, and involved at least six Tunnelling Companies. No fewer than 19 mines were exploded, with an effect that one over-excited journalist paralleled to "the last judgment." Perhaps the hyperbole was not so excessive; the St Eloi mine alone contained 95,000lb of explosives and blew a crater over 170 feet in diameter. The "Caterpillar" mine, though smaller in explosive terms, excavated a hole almost a hundred feet wider. One mine failed to go off, only to detonate almost 40 years later, in 1955.

While concrete and reinforced concrete were comparative latecomers in dugout construction, the Germans were importing large quantities of basalt and gravel from Germany into the combat zones for building works as early as 1915. The results were so effective that many of the Germans' pioneering efforts survive in place to this day. Concrete was ideal both for shelters and pill boxes but the latter were not so widespread on the British side. As Colonel E.G.L. Thurlow remarked in *The Pill Boxes of Flanders*, the idea seems to have been current that, "such works were not worth the labour or the cost, but probably the real reason was fear that a lack of offensive spirit might have been engendered if the troops had been provided with such solid defenses." It was also true that the Germans had the opportunity to build concrete pill box belts behind their existing lines and then fall back onto them as required, as was the case in the retreat to the Hindenburg Line in spring 1917. Captain F.C. Hitchcock of the 2d Leinsters got to examine similar defenses near Ypres in September 1918:

"While eating our lunch I studied the 'Pill Box' as it was the first time I had ever seen one of these famous

fortifications, and I was very interested. The name 'Pill Box' aptly describes what these 'strong points' were like. This particular 'Pill Box' resembled a miniature Martello Tower (so very common on the south coasts of England and Ireland). It stood only some six feet off the ground on deep foundations, and had a small wing extension at the back where the entrance was. Its walls of about three feet in thickness were built on a steel framework, and being made of reinforced concrete were capable of withstanding a direct hit from a 5.9. They were always impregnable to the ordinary barrage of field artillery. There was room inside for a garrison of about fifteen men. A bench had been made for a machine gun emplacement just below a narrow horizontal slit close to the ground to enable the gunners to get a wide traverse.

The Pill Box system of holding the front line in Flanders had been introduced by the German 4th Army commander – General von Arnim – in June, 1917. Our troops first came into contact with them on the Messines Ridge attack at this date, and later with often disastrous results in the third battle of Ypres. The waterlogged soil in the Ypres sector made defense of deep trenches impossible, so the problem of holding their ground was solved by the erecting of Pill Boxes. They were echeloned in depth with great skill, and always managed to stop or break up our attacking troops with either frontal or enfilading fire. Their only vulnerable point was at the entrances. A special method of attacking or dealing with these strong points had to be introduced – two parties working in conjunction. One party kept the Pill Box under heavy rifle fire and a bombardment of Mills Bombs (fired from rifles), and the other party crept round a flank to a selected point. When the latter were in position they fired a Véry light. Simultaneously both parties rose to their feet and rushed the Pill Box. Another method was to attack under cover of a smoke screen."

So it was that tactics strived to keep apace of the latest development – if not always successfully.

In any event many of the British works would be shelters rather than firing positions; often they looked very much like concrete versions of World War II Nissen Huts, unsurprisingly, since they were often lined with "elephant" shelters, and shuttered externally with sheets of corrugated iron during construction. Good examples remain at Langhof Farm between St Eloi and Bedford House cemetery, and at Warneton near Messines. During 1916 British engineers experimented with concrete as a medium, especially for observation posts, but it would be 1917, after first-hand experience of the German *Mannschafts Eisenbeton Unterstände*, or MEBU pill boxes, before the British began to use concrete widely in the front line.

On occasions raids were mounted to gather

information on constructions, as was the case in the Messines area in April 1917, when men of various Royal Engineer Field Companies were attached to infantry units for this purpose. When Messines Ridge was finally taken a detailed report entitled *German Ferro Concrete Structures on Messines Ridge and the Effect of Shell Fire on Them* was made. Nonetheless the scale of the German concrete works at Passchendaele did come as something of a shock, as also the realization that a deep belt of small but numerous and thickly scattered concrete works amounted to a new and often successful defensive system, essentially different to the old trench line. In the Ypres salient from Pilkem Ridge to Hill 60 alone more than 2,000 concrete structures had been built by the enemy. Some were recognizable as pill boxes in the sense that they had forward facing vision slits and weapon embrasures, but others were "blinds," lacking such orifices. These last were intended primarily as shelters from which the occupants would rush after the bombardment to mount their machine guns to fire over the roof, or from nearby shell holes. The whole defensive area was several miles deep and divided into three zones. The first of these was the Forward Zone consisting mainly of mutually supporting MG bunkers and wire; the next was the Main Battle Zone with its deep dugouts, strong points and trenches; the last, well back, was the Rear Zone in which the reserves were held.

Even if captured, German concrete works still presented serious problems to their new owners, the most important of which was that they faced the wrong way, with the thinnest protection and vulnerable doorways now towards the enemy. In many instances, like that of the celebrated Cheddar Villa, captured in July 1917, they would become death traps. Some of the earliest British essays in concrete work were therefore simply "turned" German pill boxes. At its simplest turning amounted to drilling a hole in what used to be the front of the work, placing a charge in it and then blowing a new entrance, while extra concrete was added to what used to be the rear, and existing doors were blocked.

New British concrete works were generally constructed of one part cement to two of sand and four of stone. Reinforcement with iron bars, expanded metal or wire ties considerably increased shell resisting capacity, as would be spelt out in the manual *Use of Ferro-Concrete in Dug-Out Construction*. Great care was necessary with the reinforcement because pieces of metal that were too large would weaken the structure by setting up lines of fracture. Properly reinforced concrete was the best medium available, as "just" 3 feet 6 inches was capable of resisting the extraordinary power of the 210mm howitzer. When building with concrete it was usual to form the work, be it a pill box or defensive shelter, by means of wood or iron shuttering, effectively building up a mould and casting the concrete within it. Where this was not possible the alternative method of precast blocks would be used.

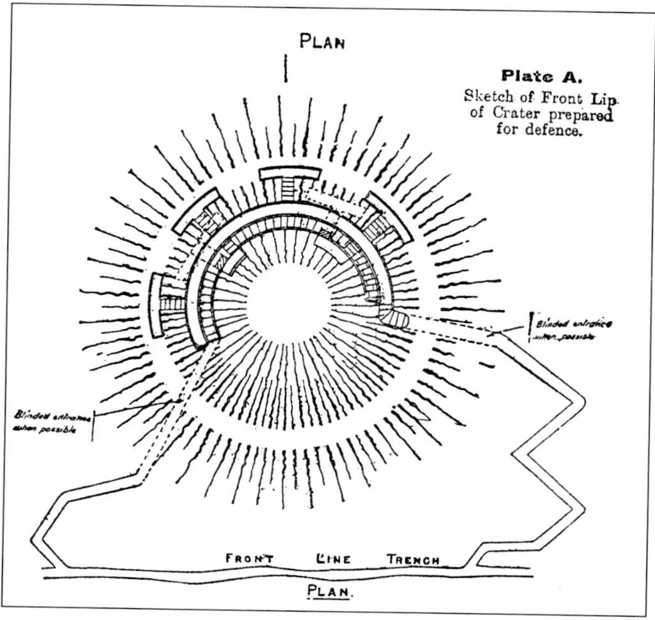

In late 1917 the British would finally begin to plan and build a system of concrete fighting works in their own front line, realising perhaps the increased likelihood of a major German offensive in 1918. One of the keys to the new plan would be a factory for precast concrete blocks and sections, with an instructional "School of Concrete" attached, at Aire-Sur-La-Lys, which ran its first courses in January 1918. A second block factory, producing a slightly different form of pill box for Second Army was also established at Arques, near St Omer. Apart from the standard First and Second Army pill boxes built from these kits, there were at least two other major prefabricated types in use in the last year of the war.

The first of these was the Moir designed by Sir Ernest Moir of the Ministry of Munitions. This was a circular concrete block wall of six feet internal diameter, with overhead cover, intended to be capable of resisting light field guns and 5.9-inch shells exploding as close as a yard away. The parts were made or gathered at Richborough in Kent and shipped from there to France. Four trucks would carry the parts to the construction site, where a team of twelve men could carry out the assembly. Perhaps a thousand Moir pill boxes were put up on the Western Front.

The second major prefabricated form, the Hobbs, was conceived by Major General Talbot Hobbs, commander of 5th Australian Division. The Hobbs was essentially of steel, being a small armor-plated cupola for a machine gunner and loader which revolved on six rollers. These pill boxes were made by the North British Locomotive Company and Beardmore's of Glasgow. About 200 would be installed on the Western Front.

Another design in use in 1918 was the simple and ingenious "ferro concrete pancake shelter," which was designed into the GHQ "stop line" behind the front, where it was easier to mix and pour concrete away from immediate sight of the German guns. This began with the excavation of a section of two foot six wide trench in a square pattern in the ground, in which one small uncut gap was left. The trench was then filled with reinforced concrete, and the square capped with a thick concrete lid. The builders now dug out the shelter, the

Far left: **The entrance to Grange Subway, Vimy Ridge. The Grange tunnel was one of the most impressive underground works at nearly 800 yards long with its own cookhouse and hospital. It was reopened to visitors in 1926 by Captain Unwin Simson. Royal Canadian Engineers.**

Left: **How to convert a crater for defense, from an American manual.**

Right: **A sentry of the Worcestershire Regiment observes through a loophole in a sandbag trench block, Villers, August 1916. He wears regulation Pattern 1908 webbing, and appears to be using a sock as breech cover for his rifle. (IWM Q 4100)**

Far right: **Army Ordnance Corps soccer players with the BEF 1917. The game was highly popular throughout the war. Soccer was played during the 1914 Christmas truce, and a battalion of the East Surreys even kicked balls ahead of them during the attack on the Somme on July 1st, 1916.**

earth having formed crude but effective shuttering. The pancake shelter was thought to be "just" proof against the 5.9-inch shell, and was intended to hide twelve men. Also built for the GHQ Line were two patterns of hexagonal pill box with two gun slits and a rear blast wall. This design would become a British standard in World War Two.

Concealment and deception were two of the best forms of defense, and by the last year of war each corps had its own camouflage officer having responsibility for the "Corps Camouflage Dump" with its stock of materials and a small factory. This latter produced bespoke items such as dummy trees, or nets cut and colored to order. Among the off the peg items in the corps store were "fish netting" garnished with knots of canvas or islands of scrim, wire netting, posts, and dummy heads and figures. These dummies filled many roles: Attracting sniper fire, deceiving the enemy into believing that a position was occupied, or that an attack was likely. Similarly available were three standard observation posts. These were the Oliver, like a small sentry box with an armored section on top; the Roland, an armored "bread bin" affair on a stretcher; and the Bee Hive, an upturned bowl-like structure with a gauze panel in the front and an irregular cover for concealment.

Sometimes it was even attempted to cover whole sections of trenches or vulnerable road or track intersections with screens. Two different philosophies could be employed, as explained in *Notes on Screens* in July 1917. Either a large and obvious screen could be used, in which case the enemy had to decide whether it really did conceal anything, and if so, how much ammunition to waste on its destruction, or the attempt could be made to conceal the screen itself. These "subtle" screens could have scenery painted on them, or be interwoven with grass or brushwood to mimic the surroundings. They were most effective as temporary cover for a battery or a small working party. Natural concealment was even more desirable where it could be obtained; folds in the ground, woods and hedge lines were obviously used, but there were more devious ploys. Sheds and haystacks could both be made to remain standing while MG positions and observation posts

were built inside. At Hussar Farm, just outside Ypres, and at some other places, the Royal Engineers employed deception on a larger scale, strongpoints being constructed inside existing farm buildings.

Dummy positions were particularly useful in absorbing punishing fire which would otherwise be directed at real posts. At their simplest these were lines of newly turned earth or dummy loopholes, sometimes with old rifle barrels projecting, or glass to reflect the light mimicking periscopes and binoculars. At the other end of the scale pill boxes were constructed of earth and then covered with cement plaster. Less permanent versions were wholly constructed of canvas and wood. Track ways and communication trenches often betrayed the presence of real posts, so it was common practice to cover the real tracks and make others to dummy positions.

Perhaps the definitive statement on camouflage came with *The Principles and Practice of Camouflage*, in March 1918. This made it clear that the objective was not merely to conceal things like trenches and posts from view, but to deceive, rendering "objects indistinguishable or unrecognisable by means of imitation or disguise." Very often camouflage did not obscure, but effectively hid by leading the observer to believe that what he saw meant something else.

Just how alien a landscape was created by the sheer human effort of exploding, digging, tunnelling, building, camouflaging, wiring, and ultimately fighting and dying, over a period years, may be readily appreciated. As one eyewitness put it in 1918:

"The country… looked as if it had been stricken by a plague, as far as the eye could see there was nothing but a sea of mud and broken trenches, barbed wire entanglements, smashed dug outs and occasionally a huge mine crater, and spread all over were steel helmets, dud shells, live bombs, discarded rifles and broken limbers."

Simply surviving in this hostile environment was likely to be a significant work of adaptation. Yet, like creatures evolving and changing to suit new circumstances, it was a challenge to which the armies would rise, and eventually overcome. As in the model of the animal kingdom, new conditions would demand new specialisms. Some of these would be specific to that environment, others would last well beyond the war for which they were designed.

Right: **German troops on maneuvers – note the grenades on the trench side.**

CHAPTER THREE
THE SPECIALISTS

If the trench environment was evolving throughout the war, then the same was true of the men in it. We have seen how the Regulars of 1914 were supplemented and diluted by Territorials and Kitchener volunteers, and finally by conscripts, but significant though this appeared to the Old Contemptibles it was probably the least important part of the process. On the battlefield the most notable difference was that minor tactical decision making devolved, first to company level, then to platoon level, and finally to section level.

Partly this was a result of the trenches. The occasions on which the commanding officer would actually see his battalion were limited, let alone have opportunity to command them in a mass. Unless involved in a major offensive battalions were likely to be split and spread, hidden across the "empty battlefield," in fire bays and shell holes, and as the war progressed they would become ever more diffuse. The enemy was even less apparent. As early as May 1915 Hesketh Pritchard, then serving as a war correspondent, was moved to remark that, though the enemy line was often in sight, "one never sees a German… beyond the fact that they use sand bags of blue and green and that there is lots of wire, I could make out nothing."

Yet devolution of responsibility was also tactical and technological; the soldier of 1914 was foremost, and almost exclusively, a rifleman. It was possible for a lieutenant colonel to give an order to his company commanders, and to know pretty well how each of the individual cogs in his battalion machine should react,

Far left: **A German post card of "our enemies" depicts British infantry in cover.**

Left: **Sniper of the London Irish Rifles, Albert, August 6th, 1918. Equipped with an SMLE with offset telescope, he wears a steel helmet and trousers cut down to make shorts. This particular daylight patrol was less than successful; of the seven taking part one was killed, and three wounded.**

and that their actions would be similar, with each man being identically armed and equipped. Their prompts were governed by orders relayed through the officers and NCOs, but the basic script was usually by the book, and the two major texts were *Infantry Training*, and *Musketry Regulations*.

Though they never became complete irrelevancies these certainties were soon under drastic revision. Specialisms and new weapons started to develop immediately. Some ideas came from enthusiastic junior officers and men in the trenches; others were developed or imposed from outside. As Hesketh Pritchard would put it, "this war... was largely a war of specialists." Eventually there would be not one or two manuals; but hundreds. Some, like *Instructions Regarding Strombos Horns*, or *Prevention of Frost Bite or Chilled Feet*, would sink without trace, but others would become akin to the ten commandments of modern war, among them slim but vital volumes like *Instructions on Bombing*, *Instructions for the Training of Platoon Officers for Offensive Action*, *The Tactical Employment of the Lewis Gun*, and *Scouting and Patrolling*. As Guy Chapman of 13th (Service) Battalion, Royal Fusiliers, would recall, "We seized and devoured every fragment of practical experience which came our way... gobbled whole the advice contained in those little buff pamphlets entitled Notes From The Front."

The increased reliance on junior leaders was compounded because battlefield communication by no means kept pace with the weapons and tactics. While platoons and sections were mobile and liberated from strict compliance to choreography by late 1918, communications and headquarters remained virtually static. Château generals who sent messages which were out of date or irrelevant before they reached the end of the headquarters drive may not have helped, but in fairness there was no reliable method of instantaneous battlefield communication. The idea of "leadership by objective," or *Führung nach Direktive*, pioneered by the German Army, eventually became widespread, not only because it seemed to achieve the best results, but because in most circumstances there were no practical alternatives. Colonels still pointed junior leaders in the right direction, gave them objectives and time tables, and shot simple signal lights which may or may not have been seen, but the era 1915–18 became very much a "soldiers' battle"; and the nature of that battle, or at least its tactics, changed from year to year, if not every few months.

Sniping

The specialism most closely allied to the rifleman's existing experience was sniping but, unlike marksmanship, sniping was a skill which was slow to gain official approval. It was an art which personalized killing, and revealed war for what it was. Sniping scarcely existed in August 1914, and this was perhaps surprising, since the verb "to snipe" had been in existence since the 18th century, and telescopic sights had been experimented with in the Napoleonic Wars. Lieutenant Colonel Davidson of the Indian Army had fitted telescopes to percussion weapons in the 1830s and 1840s, and fairly extensive use of snipers with telescopic sights had been made in the Civil War. In the South African War the Boers often used techniques which amounted to sniping, and in 1901 a sight known as Dr Commons telescopic sight had been fitted to the Lee-Enfield.

In the fall of 1914 it was the Germans who were quickest to realize the importance of sharpshooting in the developing war of position. They are thought to have deployed something like 20,000 rifles with telescopic sights by the end of the year, plus a number of sporting weapons collected by the Duke of Ratibor for army use. Sniping would never inflict the massive casualties caused by artillery or offensives, but it restricted movement, lowered morale, and denied the enemy observation and intelligence. As British tactics emerged they would stress the link between scouting and sniping, the objective being not merely to kill the enemy, but to find out what they were doing and relay that information to the forward observation officers of the artillery, machine-gun officers, and staff. Even so early attempts at sniping owed more to the skills and equipment of the big game hunter, and the enthusiasm of individual officers, than any organized system of military training. Three officers were especially influential in developing new techniques, putting them into print, and setting up the sniping schools which would become an established part of training. This trio of majors were H. Hesketh-Prichard, F.M. Crum, and T.F. Fremantle (later Lord Cottesloe).

Early sniper rifles were of two main types: Sporting guns, most of which were private purchases, and issue weapons fitted with special sights. Sporting rifles came in every conceivable caliber and type: Lieutenant L. Greener of the Royal Warwickshire Regiment (nephew of the famous gun maker W.W. Greener) carried a Ross Model 1905 0.280-inch sporter with Zeiss prismatic telescope to great effect; Crum's sniper section in the King's Royal Rifle Corps took with them a 0.416-inch Rigby rifle. Hesketh Pritchard experimented with all kinds of rifle from a Jeffrey high velocity 0.333-inch through to the most monstrous of elephant guns. One sniper section of 9th (Service) Battalion, The King's Own Yorkshire Light Infantry, boasted a mighty 0.600-inch Express rifle. This was capable of smashing armored shields, but had the disadvantage that firing prone could crack a collar bone with its harsh recoil. Among the service pattern weapons the SMLE Mark III was most commonly available, but there were other options. Some used the older "Charger Loading" or "Long" Lee-Enfield; at the end of the war others opted for the American-made Remington-Enfield P14. A few lucky characters managed to obtain captured German equipments, usually the Gewehr 98, fitted with Zeiss, Goerz, or Luxor telescopes.

Above: **The Symien Sniper Suit of painted scrim, as depicted in a camouflage manual of 1918.**

Above right: **Men of US 77th Division receive instruction on camouflage from the British, Moule, May 1918. Note the sniper, foreground left, in coveralls and foliage. (IWM Q 10316)**

The Canadians, whose forces started the war with the Ross rifle, also used it widely for sniping. It was accurate, but as a general issue too delicate for the hurly burly and filth of the trenches. There were horror stories

of incorrectly fitted bolts flying out into the faces of their users, but authenticated instances of this accident were few and far between. More common was simple failure to function. In early 1915 Canadian battalions were equipped with four Ross rifles with telescopic sights, usually of the American Warner and Swasey prismatic type.

Lack of optical equipment was a problem from the outset as the industry was dominated by German manufacturers, and until the summer of 1915 the army would be largely dependent on individual private purchases and small orders from gun makers. Up to July 1915 1,260 government orders were placed for telescopic sights and the list of suppliers was a veritable directory of top-class sporting gun makers including Purdy, Holland and Holland, Churchill, Lancaster, Rigby, Westley Richards, and Jeffrey. These scopes were expensive pieces of equipment costing anything between £6 and £13 ($30–$65 at 1914 exchange rates), at a time when the SMLE cost £3 10 shillings (£3.50), and a Mills bomb just 5 shillings (£0.25). When not actually fitted to the rifle these sights were usually carried in leather cases with buckle down flaps on a sling over the shoulder.

Additionally numbers of magnifying or "Galilean" sights were employed. As Fremantle described them they consisted of a convex and a concave lens, in which the target was seen the right way up. The lenses were not linked but attached individually to the gun at muzzle and breech. They had a narrow field of view and it was not possible to fit any cross hairs, but they did have compensations. A set could be purchased for between 10 shillings (£0.50) and £5, they required little or no setting up, and were easy to use. Of the four main types the Lattey optical sight was most widely used, and consisted of two simple lenses. The Neill sight, also known as the Barnet or Ulster, was offset to the left, allowing the ordinary iron sight to be used if desired. The sights of Martin and Gibbs were more complex, featuring an adjustable rear aperture.

It was not until May 1915 that an official specification for the fitting of telescopes to service rifles was finally approved, and even then it was framed in the most general terms to allow the use of available stocks. Only in 1916 was anything like mass production of telescopic sights for the SMLE achieved. Over half now came from the Periscopic Prism Company of Kentish

Town, with Aldis of Birmingham the second largest producer, and Winchester of New Haven, Connecticut, the third. These three manufacturers between them produced over 90% of the 10,000 telescopic sights provided for SMLEs by the end of the war.

Strangely the majority of these sights were not mounted centrally over the barrel but offset to the left, allowing the rifle to be loaded with the normal five round clips. This had several disadvantages, as was later explained by Hesketh-Pritchard in his book *Sniping in France*, and repeated in *The Field* magazine

"This caused all kinds of errors. The set-off, of course, affected the shooting of the rifle, and had to be allowed for. The clumsy position of the sight was apt to cause men to cant their rifles, and some used the left eye. Worse than all, perhaps, in trench warfare was the fact that with the Government pattern of telescopic sight, which was set on the side of the rifle, it was impossible to see through the loopholes of the plates which were issued, as these loopholes were naturally narrow."

To get into the right position many men had to resort to screwing an extra piece of wood to the rifle butt, sticking a shell dressing to the woodwork to act as a cheek piece, or resting the head against an ammunition bandolier. The offset scope was in any case based on dubious logic. Charger loading was critical to rapid fire, but this was seldom, if ever, needed by the sniper, who was in the business of single, aimed, and unexpected shots. Each shot might of course be taken rapidly at a fleeting target, but the victim was unlikely to stand around waiting for the full magazine. Moreover snipers naturally preferred not to shoot more than once for fear of revealing their position.

Complaints led to more scopes being mounted in the top position but only in 1917 was a completely new layout devised for use with the P14 rifle, inspired by the capture of a German Hensoldt telescope with adjustable fittings. The complete new British assembly, approved in April 1918, was known as the "Rifle, Magazine .303in Pattern 1914, Mk 1* (W) T." The "W" stood for Winchester, as the Winchester-made P14 had always been found most reliable and was often used in the making up of sniper sets. The scope itself was known as the Model 1918, and was made by Aldis.

Observation equipment was the sniper's other weapon, used both to locate the enemy for sniping, and to provide information. Stalkers' and signallers' telescopes were widely favored for a number of reasons. The normal box periscope of the sentry was generally too bulky though it gave a good field of view. Binoculars were sometimes used but presented a direct and relatively broad target to the enemy. The telescope, by contrast, was already the preferred choice of the game and target shooter. It gave little for the enemy to see and could be pushed between sandbags, or through narrow gaps, or even left set up in a steady rest. Many were provided with a pull-out shade which kept direct sunlight off the lens and prevented tell-tale reflections. When brought rapidly into use Fremantle recommended that the shade be opened first and the telescope itself be slid smoothly to prearranged settings which could even be scratched onto the outside of the

Right: **Men of the York and Lancaster Regiment near Roclincourt, January 1918. A variety of camouflaged robes and "crawling suits" are worn. (IWM Q 23580)**

Far right: **Lewis gunners and snipers of the Queen's (Royal West Surrey) Regiment pose with cuddly toys, *c.*1918. The fleur-de-lis of the sniper is worn on the shoulder; the LG in wreath of the Lewis gunner on the left cuff.**

tube. Even so the observer had to be extremely careful and keep his wits about him; as was discovered by sniper F.A.J. Taylor of the 2d Battalion, The Worcestershire Regiment, when he peered from a well prepared hide near Ypres in 1918.

"I was gaining experience in using the telescope when one day I became suspicious of a small dark, triangular patch in a mound of earth about three hundred yards in front. Staring at it for some time I couldn't make out what was a light circular patch in the middle of it. When, to my utter amazement I saw very clearly the side of the face of a German, with light brown hair. His profile showed as he turned away. I kept watching as I speculated, then suddenly I saw a whisp of smoke and instinctively ducked as his bullet zipped viciously through the sandbags over my head."

An officer appeared on the scene and attempted to snipe back at the German, but it was only when artillery could be brought to bear that he was dislodged.

The official number of snipers per battalion was eight, though many experts preferred more. Crum, writing in 1916, recommended an establishment of 16 men, a sergeant, a corporal and an officer, and thought that in certain circumstances as many as 24 might be required. Taylor's section in the Worcestershire's was at least 13 in strength at the end of the war, and included a sergeant and a corporal, and this was by no means unusual. 124th Infantry Brigade may have been keener than most on sniping, but towards the end of the war it specified a sniper organization, per battalion, of no less than one officer and 64 other ranks. Though sniping was a dangerous occupation it did have attractions. Snipers were usually excused fatigues and were not normally employed at night. Though they had a good deal of special equipment to clean and carry, it was not as arduous as duty with machine guns or working parties.

Occasionally men worked as loners but standard practice was to operate in pairs, one man observing with the telescope or periscope, the other ready with the rifle. Where possible these pairs were spread across the battalion front to give greatest coverage and were relieved at intervals by fresh men. Sniping NCOs were allotted sectors of this front to oversee by the battalion officer, who was known variously as the sniping or intelligence officer. Within each sniper pair one man usually specialized in the shooting and the other in observation, but when the concentration of the latter lapsed it was better to swap duties for a while . It was desirable that the pairs chose their own partners, as personality clashes or misunderstandings could have

fatal results. Depending on the cover available the sniper could fire from virtually any posture, but prone often presented the least target and the best chance of blending with the surroundings. Many used the "Hawkins" position in which the toe of the rifle butt actually rested on the ground, giving a steady aim and less stress on the rifleman. Sandbags, slings, rests and depressions in the ground could all be used to improve accuracy.

When a shot was taken the sniper kept a few simple points in mind. First, polished technique was not as important as consistency; provided he did the same thing every time he would be accurate. The same was true of equipment and ammunition, best policy was sticking to the same rifle, and the same cartridges – clean and corrosion free. Next, when a target was located the sniper would look in the general direction, then bring the gun up to the target area. Searching through the rifle scope was all right as a supplement to the observer's work, but not the way to take a quick shot. Lastly the firer would have to be conversant with the principles of aiming off. Many telescopic sights were fixed, but even when they were not, there was seldom time to adjust them for wind or a moving target. The rifleman therefore had to judge wind speed and movement when aiming, allow for cross winds, and aim off, ahead of movement. These allowances were sometimes quite large. At 300 yards Fremantle calculated it was necessary to aim off between one and three feet depending on the strength of the wind. A Mark VII 0.303-inch bullet travelled 200 yards in a quarter of a second; a walking man travels at about six feet per second so it required a shot 18 inches ahead of him to secure a hit at that distance. These factors multiplied with range, so a running man at 600 yards needed a correction of 17 feet! Clearly this was hardly ever going to be successful so few snipers bothered with moving targets over 300 yards. Contrary to popular opinion there was little work done at very long ranges, and many sights were left set at 200 yards.

Whether snipers should regularly venture into no-man's-land was a matter of debate. To do so brought the gun closer to the target, often from an unexpected angle, and many good tallies were achieved in this way, yet there were also severe disadvantages, as Canadian sniper Herbert McBride pointed out:

"When it comes to crawling alone out in front of your own trenches… I am off that stunt also – by preference that is – though I have done quite a bit of it. Here, a man is strictly 'on his own' and is pretty much apt to be up against it if anything goes wrong. His field of fire is much limited, no moving about can be indulged in, and generally but one or two shots may be fired before the show is over for the day. Then comes the long, fearful wait until darkness sets in, before the crawl back to your own trenches may be begun."

An early exponent of the "crawling out" technique was Captain J. Grenfell of the 1st Royal Dragoons, who attempted it in November 1914:

"I took about 30 minutes to do 30 yards; then I saw the Hun trench, and I waited there a long time, but could see or hear nothing. It was about 10 yards from me. Then I heard some Germans talking, and saw one put his head up over some bushes, about 10 yards behind the trench. I could not get a shot at him, I was too low down, and of course I could not get up. So I crawled on again very slowly to the parapet of their trench. I peered through their loophole and saw nobody in their trench. Then the German behind me put up his head again. He was laughing and talking. I saw his teeth glistening against my foresight, and I pulled the trigger very slowly. He just grunted and crumpled up."

If one did have to venture out beyond the line the best aid was a "sniper's robe" or "crawling suit" together with a camouflaged canvas cover for the rifle. These robes and suits came in a variety of patterns, many of which were made, or at least painted, near the front. The earliest versions were often cut like a loose coat, with or without a hood. Usually of canvas, they could be painted with disruptive splodges of colored paint to break up the outline of the body. The coat design was found to impede crawling so later versions were made like baggy coveralls.

By the end of the war corps Camouflage dumps were concentrating on two moderately standard types, the "Boiler Suit" (coverall) pattern, with detached scrim hood, rifle cover and gloves, and the very modern looking Symien Sniper Suit of painted scrim. This was basically a long loose-fitting jacket with an integral hood and separate legs or pants. It was teamed with gloves and rifle cover. Other artistic touches could be added with local foliage, straw, and vegetable stalks as appropriate to the surroundings. A simple alternative hood could be made by pulling a sandbag over the head and separating the fibres a little to create vision slots. Among trees a domino or speckled hooded cape was sometimes worn. The key always was to blend into the background, and not present a definite outline or shadow. The sharp outline of the service cap was best avoided, or at the very least the wire stiffener could be removed to soften it. Sometimes, as Crum relates, the outline of the head was destroyed entirely with masks, painted to represent brick or stone or covered with gauze.

Many snipers remained in or around their own lines, often in posts connected by short saps. It was important that there be a number of posts, well concealed and protected, so that the teams could move from one to another obtaining new and unexpected fields of fire. Many positions were burrowed through the front parapet, usually at night. Protection was improved by steel loops or sandbags filled with stones or scrap metal, but these had to be used sparingly as they were a distinct liability in the case of bombardment. Roughly painted patches of dark material on the trench parapet helped provide an illusion of irregularity, making it more difficult to spot the sniper. As far as possible real sniping positions were at an angle to the front making detection more difficult and enfilade fire possible.

Where sniping was well organized the battalion Sniping Officer kept a plan of the locations of the posts, and ordinary infantrymen were kept out of them to avoid compromise. Range cards were maintained in the posts, or on the curtain behind the loop, detailing distances and locations of enemy positions, and

Left: **Officer with bullet-riddled metal loop hole plate. (IWM Q120)**

topographical features. Particular care was taken to avoid smoking or cooking near a post, and the firers would attempt to conceal muzzle flash from rifles by screening or shooting from just inside the aperture. Dust which could be raised by muzzle blast was best avoided, or damped down with a little water.

One of the most important duties was to locate and eliminate snipers on the other side of the line. Many ruses were used, designed mainly to get the enemy sharpshooter to give away his position. In many cases two or three British snipers would work together. One of the team would provide the decoy, cautiously opening a loophole with a stick, showing a dummy head, or firing a few rounds, and when the German engaged him the rest of the team would come into play from different angles. Some were too cautious to be taken in by such simple ploys, and their lairs had to be painstakingly located over a period of days. Hesketh-Pritchard became an expert in examining the locations where enemy snipers had been active and judging their positions by the angles of bullet strikes on trees and sandbags. One favorite device was the dummy head, which could be made especially realistic with a cigarette between its lips, smoked from a distance, by means of a tube, by the sniper. When the head was struck it could be lowered and examined to determine the path of the shot.

When enemy snipers were quiescent, enemy officers were a tempting target, and distinctions in dress helped identify them. Hesketh-Pritchard recommended that when snipers were operating in defense, perhaps against an enemy offensive, they should be given a definite battle station and tasked to pick off leaders, machine gunners, and trench mortar teams. The Germans naturally responded in kind and, despite Hesketh-Pritchard's warnings, many a British officer who insisted on wearing riding breeches was picked off by an alert German sniper. As early as December 1914 it was realized that the Sam Browne officers' leather cross belt was attracting unwelcome attention, and so a routine order was issued instructing officers to adopt the same equipment as their men. Many, however, ignored the order, and some no doubt died as a result. High legged private purchase trench boots, and wrist cuff rank insignia could also prove liabilities, and gradually these too were discarded.

Armored enemy loops presented a particular problem since, although the standard 0.303-inch round would "ring the bell" when it hit such a target, it could not penetrate at anything but the closest range. Short of calling down the artillery the best solution was a high velocity sporting rifle, or large caliber elephant gun. When the enemy tried to knock out British loops in similar fashion, the reply was the double loop. One steel plate was placed to the front in the ordinary way and concealed. To the rear of this was placed a second loop, ideally in a sliding rail. Loops were also constructed of two plates with the gap between packed with earth.

In the early stages of the war sniping was slow to gain official approval, or the facilities for training which befitted such an important skill. This was probably both because of its "dirty" reputation, and because of the high degree of individualism it entailed. Sniping eventually developed officially as a branch of scouting and intelligence work. Training was sanctioned at army and corps level in 1916, when the major schools were established. Even then each establishment had its own curriculum and strengths. The principal aim of First Army School at Linghem was to turn out qualified instructors for corps schools, brigades and battalions. Second Army School at Acq had its own reconstruction of a German trench system in a quarry, and placed a strong emphasis on the scouting aspects. Night training was also given with the men groping about in specially devised dark glasses watched by instructors. Fourth Army School was against the steep chalk slope at Bouchon and was equipped with prefabricated buildings made by the engineers at Amiens. General, and surprisingly advanced instructions appeared as a part of Army Printing and Stationery Service Manual 195, *Scouting and Patrolling*, 1917.

Battalion snipers could be trained virtually anywhere and usually received a course lasting about two weeks from the intelligence officer and senior NCOs, after which they were entitled to a distinctive badge. There were several patterns but later in the war the standard model was a brass or cloth fleur-de-lis worn on the right upper arm. The course was both practical and theoretical. Surviving syllabuses show there was usually a lecture in the morning, on observation, map reading, or telescopes; in the afternoon there would be a practical session. These included not only rifle shooting but unarmed combat and lessons in observation. Hesketh-Pritchard emphasized that the objective was not merely to produce accurate shots but quick shots:

"After finding out errors in the ordinary way by grouping, we eschewed as far as possible shooting at targets; the round black bull on the white ground was very rarely used, and all kinds of marks were put up in its place. The head and shoulders was the most efficacious target, and practice was further on at dummy heads carried at walking pace along trenches. In fact, where such appliances as we had at the school were lacking, it was far better to allow snipers to shoot at tins stuck up on sticks than to permit them to become pottering target shots."

Exercises could be enlivened by watching an expert snipe from posts, or a trench system, and getting the men to work out where the shots came from; attempting to spot a camouflaged instructor; or patrolling across a piece of ground until spotted and chased off by an enemy. The most remarkable thing to the modern observer of this training was just how advanced it could be. The techniques of camouflage, shooting, movement and unarmed combat set forth by the enthusiasts of 1918 would be little out of place in the modern battle school.

Grenades and Bombing

If the role of some riflemen was transformed in becoming snipers, many more would be changed by becoming specialist "Bombers." Eventually issues of grenades would become universal, giving the soldier a permanent second weapon. Prior to the war grenades were virtually obsolete in the British Army; use by the Russians and Japanese in the war of 1904–05 had caused a ripple of interest, but until the outbreak of the Great War it was envisioned that bombs would only be used by the Royal Engineers. The

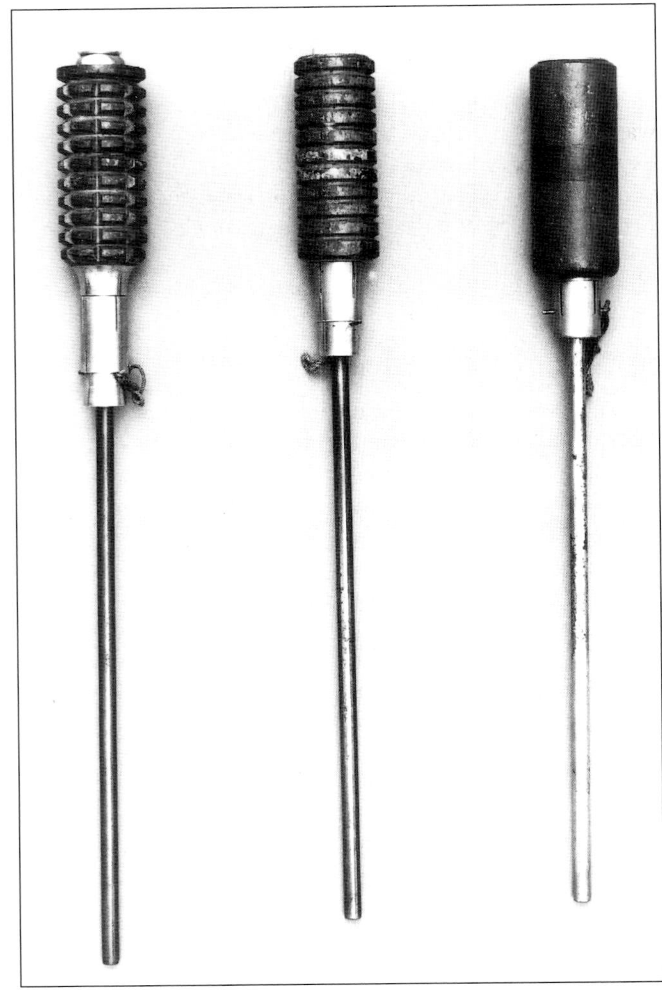

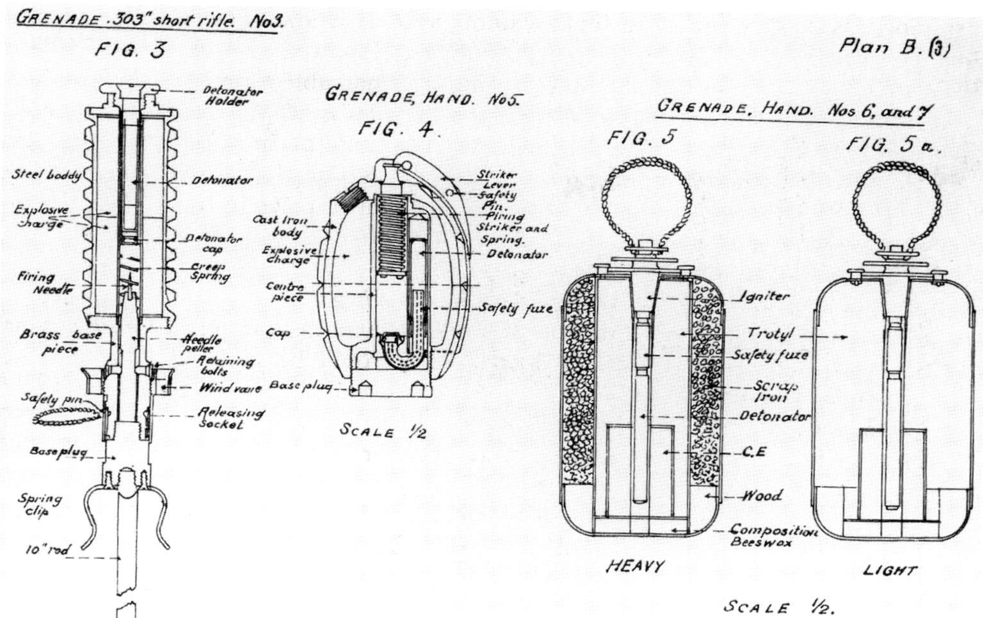

Far left: **Hale type rodded rifle grenades, the No 20 Mk 1; the No 24 Mk 1; and the smooth bodied No 35.**

Left: **Diagrams of British grenade types, 1915: Left, the Hale No 3; center the No 5 Mills bomb; right the Nos 6 and 7 Emergency type Lemon grenades.**

only grenade issued to them was the No 1. To modern eyes this is a peculiar piece of equipment. Mounted on a cane handle, it had a cylindrical brass head, around which was a fragmentation ring divided into segments. It exploded on impact, and to ensure it landed nose first it was fitted with a long webbing streamer. *Musketry Regulations* gave the throwing instructions as follows:

> *"The tail is unwound and allowed to hang loose at full length… the cap is turned from "travel" to the "fire" position… the safety pin is withdrawn… the grenade is thrown by means of the cane… the latter is grasped between the end furthest from the grenade itself and the attached point of the tail, i.e. on the grooved portion. The grenade is thrown in the required direction either under or overhand, care being taken that the tail cannot entangle itself with the thrower or with any object near him."*

The thing worked, but was expensive to make, used a good deal of raw materials, and was next to impossible to mass produce. From the thrower's point of view it also had the drawback that if it was struck on anything once armed it would explode, and in the confines of a trench this was all too likely. Despite placing orders with outside firms, and design changes including a shortened handle there were never enough No 1 grenades to meet anything but a fraction of the demand. At the outbreak of war the only other manufacturer of grenades in Britain apart from the government Royal Laboratory had been the Cotton Powder company of Faversham in Kent, which had been engaged in the manufacture of another percussion or "explode on impact" grenade for export to Mexico. Supplies of this were bought in for British Army use as the No 2, but again it was too few too late.

The proprietor of the Cotton Powder company was inventor Frederick Marten Hale and one of his new devices was a bomb which could be shot from the muzzle of a rifle by means of a metal rod. Two versions, numbered 3 and 4 were adopted by the British government. No tactical doctrine yet existed for the rifle grenade's use, but as a device which would fly further than a hand-thrown grenade, but could be used from closer in than artillery, it was quickly embraced as being of very useful harassing value.

In the meantime trench warfare was consuming ever increasing numbers of bombs. By November 1914 about 1,000 were being issued each week, while Sir John French stated that the minimum requirement was 6,000. The next month supply had reached 4,100, but the stated demand leaped to 10,000 per week. The troops, and particularly the Engineers, having nothing to winkle out the better provided opposition from their pits and trenches, began to take matters into their own hands and manufacture their own crude substitutes. As one ditty of the time put it:

> *"All things bright and beautiful*
> *Gadgets great and small,*
> *Bombs, grenades and duck boards*
> *The sapper makes them all."*

Above: **Men of 2d Battalion Argyll and Sutherland Highlanders in the trenches on the Bois Grenier sector, 1915. The grenades are No 2 Mexican stick types, and a No 8 or 9 Cylinder.**

Right: **The British No 15 ball grenade. Due to massive demand for bombs the No 15 would remain in use through the battle of Loos despite its old fashioned design.**

Far right: **The fragments generated by the explosion of a Hale type rodded rifle grenade. With a range of about 160 yards, these missiles burst into upwards of 175 wounding shards.**

Best known of these local improvisations was the "Jam Tin Bomb," known to some wits as "Tickler's Artillery," after one of the manufacturers of jam supplied to the front. This weapon was childishly simple, consisting of a tin can, containing explosive and a detonator, initiated by a length of fuse. The compilers of *Notes From The Front* caught on quickly, and explained to budding bomb makers:

"Very careful experiments should be made to ascertain the length of safety fuse to give best results. One inch of fuse should burn for 1¼ seconds, but each lot of fuse should be tested. Cavalry and Infantry should be practised in the making up of these bombs, using pebbles for the missiles, dummy wooden gun cotton, and drill or dummy detonators.

When some expertness has been attained, a powder flask should be substituted for the gun cotton and straw or sand used to represent the iron missiles. Men should have some practice in throwing the various types (made up 'dummy') from out of one trench into another without exposing themselves. They must be alert and throw immediately the fuse is lit."

The 2d Battalion of the Black Watch was one of the most organized producers of Jam Tins, and helped set up a small factory complete with red danger signal on the outskirts of Le Plantin, under the watchful eye of Major Smith, Royal Artillery. Technique in use varied slightly from unit to unit, and a pipe or cigarette often proved handy to light the fuse. One of the most dramatic accounts of a close range fight with such bombs comes from the memoirs of Sir Philip Neame, who, in December 1914 as a Lieutenant of the Royal Engineers near Neuve Chapelle, had to stem a German bombing attack:

"I ran forward and asked what was going on. The first answer was from the Germans, for a black object the size of a cricket ball came sailing through the air, landed in the trench behind us and burst with a terrific bang and the whine of whirling bits of metal. The sergeant then told me that he was the bombing sergeant of the West Yorks and that the two men with him were all that was left of his bombing squad, the rest having been killed and wounded. He said that the German's bombs out ranged our own, that our bombs were 'duds' and that he could not get them lighted.

We were interrupted by a perfect fusillade of bombs, this time coming from at least two directions, some of which landed in the trench and some on the parapet; one of the men with us was wounded. I looked round, wondering what to do, with, for a moment a helpless blank in my mind."

Having been given a bag of Jam Tins, Neame prepared to advance. There were no lengths of safety fuse available to light the bomb fuses from, and he was not a smoker, so he was reduced to using matches. According to him an ordinary flame from the match would not work, but the hot glow as the match was struck would. He therefore held the match head tight against the bomb fuse and rubbed the matchbox across it:

"I stood up on the fire step, head and shoulders showing clear above the trench, took a quick look towards the Germans and threw my first bomb. It hit the parapet near where I could see the Germans and exploded with a roar. Our bombs, though heavy to throw, were very violent and destructive. I crouched down and quickly prepared another bomb. As I stepped down a rifle bullet cracked past close to my head and a fraction of a second later there was the stutter of a machine gun. From then on every time I showed my head and shoulders a rifle or two cracked and the machine gun fired. I soon realised that the machine gunner was a little slow in his reactions and always just too late, but why I was not hit by a bullet I cannot imagine. While I crouched a German bomb came over, fell right in the crowded trench just behind me and burst with a frightful crash, killing and wounding many of our men. They cleared back out of that bay of the trench and I withdrew once more behind the next traverse. That was the last bay I gave up; it was here that I held the enemy. The trench all round me was by this time a dreadful sight, men lying with bodies broken and maimed. A bomb inflicts terrible wounds. I never saw anything worse in the succeeding years of warfare."

So the duel continued, Neame ducking up and down throwing bombs supplied from the rear. Behind him a couple of men set about building a block across the trench. On one occasion a fragment blew his hat off. So they continued "shying coconuts" at one another, until the Germans had had enough. Apart from the horror and bravery, it is apparent from this, and other accounts, that combat in the trenches, even in late

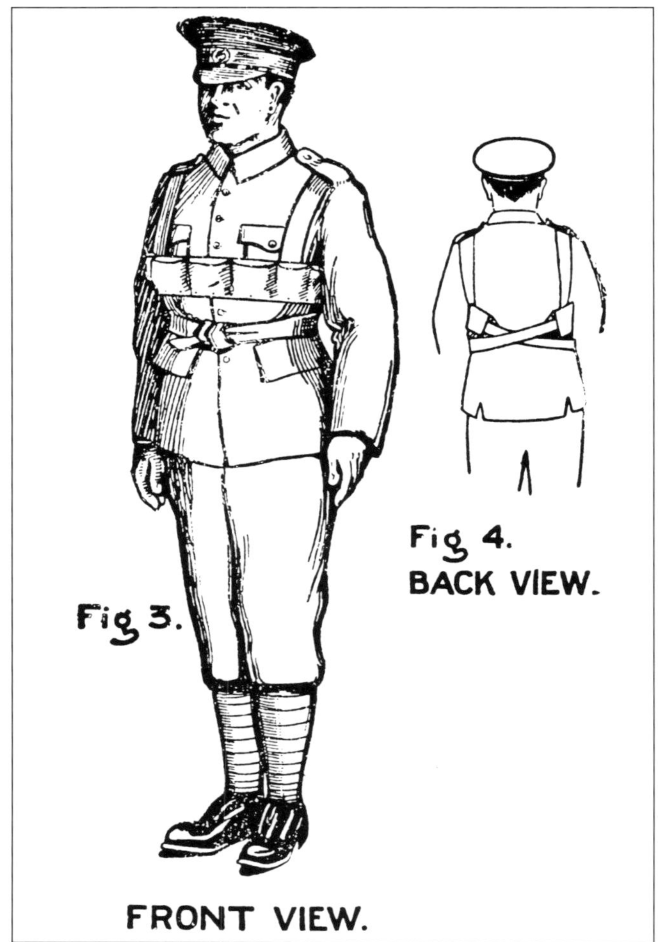

1914, was no longer a matter of companies and skirmish lines. The new environment called for new weapons and new tactics, and war was now being fought section to section, bombing squad to bombing squad, even man to man.

Although the Jam Tin obtained widespread currency it was not the only bomb improvised in France in the first six months of war. Others included the hair brush or racket bomb, literally a square of explosive wired to a handle; another was the Battye bomb. The Battye was named after its inventor, Major Basil Condon Battye of the Sappers and Miners, Indian Army. His bombs were simple castings produced in Army workshops at Bethune. Battye went on to theorize on tactics, and some of his writings were later absorbed into official publications. Yet another new grenade was the work of Lieutenant Colonel Henry Newton, a pear shaped device, known tongue in cheek, as "Newton's Pippin," which would also subsequently appear as a rifle projected bomb.

At home Colonel Jackson's infant official Trench Warfare Section of the army staff was also doing what it could to organize the production of Emergency Pattern bombs. Grenades 6 and 7 were simple tin cylinders colloquially christened Lemons, and bombs numbers 8 and 9 were double cylinder types, a slightly more professional version of the Jam Tin. The No 12 or Hairbrush similarly was nothing more than a factory version of the old racket bomb. Perhaps most vilified in this collection of unusual stopgaps were the Pitcher grenades, numbered 13, and 14, in the official series. So notorious were they that the official *History of the Ministry of Munitions* admitted that, "accidents with them were so numerous that they won for bombers the name of 'The Suicide Club.'" One of these accidents was reported by Brigadier General H.C. Lowther, commanding the 1st Guards Brigade in August 1915:

"At about 3pm Corporal Holden of the 1st Battalion Coldstream Guards, who was the only man in that battalion

Far left: **One of the first "waistcoat" (vest) grenade carriers, as illustrated in the manuals.**

Left: **Contemporary sketch showing British bombers clad in six-pocket grenade vests.**

trained in throwing the Pitcher bomb threw three of these bombs. Two failed to explode. The caps of all these bombs have been difficult to remove. Corporal Holden then took up a fourth bomb and the cap of this was even stiffer than the remainder had been. Just as he succeeded in getting the cap off and the string came away, the bomb exploded in his hand, killing him outright."

Defensive precautions developed as did the bomb. Trenches themselves were cut as square at the top as was practical, rather than sloping inwards, and a row of sandbags was sometimes laid to stop grenades rolling or tumbling inside. Some trenches were protected with wire mesh set up at an angle, and so positioned that a grenade which passed over it must also pass over the trench behind.

A refinement of the use of netting was to put a wide mesh over the most vulnerable sap heads and craters, most likely to be taken by the enemy – so that such points could be bombed into by British troops, but not bombed out of by the enemy whose throwing arms would be impeded. In all cases it was stressed that netting was not to interfere with the free use of rifles and fixed bayonets by the occupants. Grenades in defense were generally used where the ground to the front of the position precluded direct fire with rifles and machine guns. Bombing posts would also be established in trenches and dugouts behind the main trench line, so that in the event of capture the enemy could be rapidly counter-attacked and bombed out again.

It was during late 1915 that bombing in the trenches came of age, both in terms of achieving an adequate supply of reliable equipment and the evolution of new tactics. The year 1915 also saw the advent of the bomb which would, with various minor modifications, see the British Army through the next 60 years, the No 5 or Mills Bomb. Designed by William Mills of Sunderland the new bomb took as inspiration a Belgian device patented by Captain Leon Roland in 1913. This Roland bomb was segmented both internally and externally, and had a spring-loaded striker held in place by a lever until throwing. Mills had disagreements with Roland's agents, and during the early part of the war improvements to the design were registered in the Englishman's name, thus creating a legal and financial tangle which was not sorted out until 1917. The new Mills bomb was however an instant success; the first trial batch reached France in March 1915, and a significant order followed in April. Soon the troops at the front were dividing whatever stocks they had into two categories, "First Class Bombs" which really meant the Mills, and "Second Class Bombs," which covered the emergency types and French grenades.

In official circles there was some reluctance to accept the Mills. It was neither designed as an offensive or defensive grenade, although it had some of the properties of the latter, and it was time fused. The time fuse was reminiscent of the emergency bombs, and in theory it might have been possible to throw back. A constantly compressed spring was also seen as a shortcoming. Empirically none of this seemed to matter very much. The Mills was easy to use, relatively safe, powerful, and the troops liked it. Throwing procedure was simple indeed: The bomb was held in one hand with the lever securely gripped down, the safety pin was pulled, and the bomb thrown. As it left the hand the lever flew off, and the striker snapped down initiating the time fuse. When the fuse reached the detonator the grenade went off.

One Bombing Officer conducted trials to determine just how effective the Mills bomb was. He discovered that landing ten yards from its target a hit was certain. At 20 yards there was "some chance" of disabling the opposition, but at 20 to 25 the chances were four to one against. At 25 to 30 yards the chances of safety were ten to one in favor, and one was practically safe at 35 yards in the open, though a prudent bomber was still wise to throw himself flat, or take cover, as occasionally larger fragments of his own grenade would fly further. Demand for bombs continued to spiral, and so it would be the end of 1915 before there were enough Mills bombs for all. Many would have to fight at Loos in the fall of that year with Second Class Bombs, especially the baseball-like No 15.

Tactically, early 1915 was particularly significant for the grenadier, with the evolution of officially sanctioned Trench Storming parties. Interestingly, although the idea was promulgated by manuals, its genesis appears to have come from the front line, one of many such instances where the war, at least in terms of technology and minor tactics, appeared to be driving itself. The early Storming Parties consisted of bayonet men to winkle out the opposition with rifle and bayonet, grenadiers, carriers, and sandbag men to block the trench when the furthest point was reached, and to cover side entrances. In the assault grenadiers would throw over the trench traverses, and then bayonet men would burst around the corner to ensure that all was clear before the party carried on to the next traverse. With the tactics came new impedimenta, bandoliers, belts and boxes in which the stormers would carry their munitions. These would be followed by grenade "waistcoats" (vests), usually with six or ten pockets depending on which bombs were carried. A bomb bucket also came into use.

Training was quicker off the mark than with sniping. A universal system with formally laid down tactical objectives soon became available to all units. This recommended that all officers and 12 NCOs in every company throughout the army be trained with live grenades. They could then pass on the techniques within their commands. Before live training the men were to be instructed on the construction and action of the grenade, the properties of fuses, and the making up and firing of small charges. After this as much practice as possible with live bombs was encouraged, though it was stated that, "familiarity with explosives must not be allowed to induce carelessness in handling them."

By the time that the manual *The Training and Employment of Grenadiers* was published in October 1915 the role of the grenade was already under revision;

Right: **US 369th Infantry Regiment, 93d Division, outside Mouffrecourt, 5 May 1918. Note VB rifle grenade on weapon in foreground.**

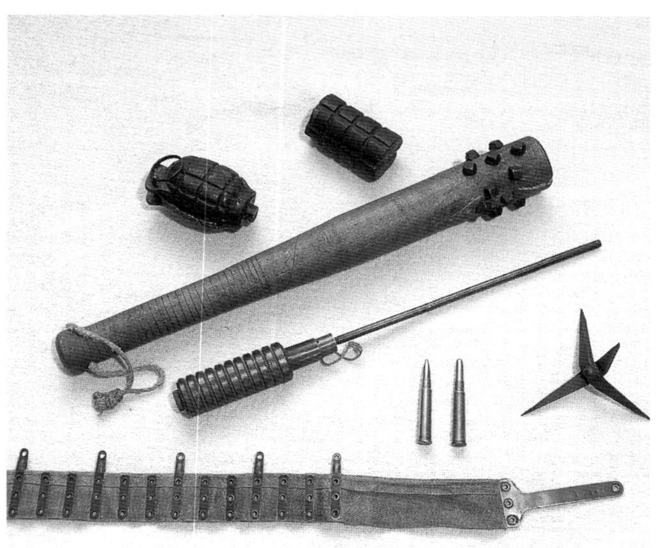

Below left: **A metal studded, lead weighted, trench club, surrounded by a Mills No 23, Battye hand grenade, Hale rifle grenade, a caltrap designed to injure the feet of unwary horses and troops, and a Vickers machine-gun belt, dated 1914. The two rounds of 0.303-inch ammunition show both the pointed bullet current during the war, and the rounded projectile of the obsolete cartridge.**

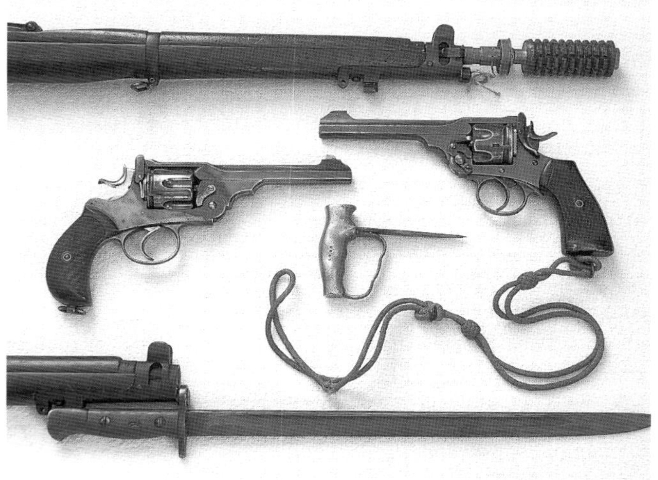

Below right: **The Webley WG Army Model revolver, left; and right the Webley Mark VI. The SMLE rifles are fitted with top, the Hale No 3 rifle grenade, and bottom the 1907 Pattern sword bayonet.**

"The nature of operations in the present campaign has developed the employment of rifle and hand grenades both in attack and in defense to such an extent that the grenade has become one of the principal weapons of trench warfare. Every infantry soldier must therefore be given instruction in grenade throwing."

Within each platoon a nucleus of one NCO and eight men was now to be formed with a higher degree of training, either to work with their platoon or provide a reserve for special operations. In the cavalry this sub-unit was to be an NCO and four men per troop. All such men were to be selected from "the very best, bravest and steadiest in emergency," with special preference for tall men with a fondness for outdoor games. It was a statement which summed up the ideal attributes of the grenadier over the centuries and, like his historical forefathers, the Great War grenadier was felt to be entitled to a special mark. In this case it was a badge of proficiency, usually a red flaming grenade, sewn to the upper right arm. The Bombing Officer – one subaltern per battalion, who was not supposed to lead the bombers in action, but to supervise training and organize supplies – had a similar badge, but with white flames. Occasionally there were regimental variations on this scheme with slightly varying types being purchased out of regimental funds or made to order. Some wore an engineer or artillery grenade collar dog as an arm badge; the Liverpool Scottish had an impressive flaming grenade of their own in brass.

A major revision in training procedures took place in late 1915, with greater emphasis on practice and throwing techniques. After overcoming "the natural fear of the grenade itself" the next step was to develop accurate throwing and distance to outrange the enemy. Normal throwing was overarm, but for shorter distances grenades could be "lobbed from the shoulder by an action similar to that employed in putting the weight." Throwing time fuse bombs like the Mills was deemed rather less hazardous as normally there was no ill effect if they were banged or dropped. Stick types could be thrown like a dart. Practice throws were made standing, kneeling and prone.

In the offensive the basic bombing unit was now christened the grenadier party. Its ideal composition was an NCO, two bayonet men, two throwers, two carriers and two spare men. The bayonet men were to be "quick shots and good bayonet fighters," and their duty was to protect the throwers, "at all costs." The most usual tactics of the group were already familiar: To bomb along enemy held trench, halting and spreading out at every traverse, after which the bombers would throw, and then the bayonet men would rush around the corner. In preparation for the attack the following distributions were recommended:

"a) Grenades on the scale of three grenades per man, to be issued in bulk to those units detailed to open the attack, the issue being made in sufficient time for them to be distributed as required.
b) Small depots to be established at frequent intervals along trenches from which the attacking columns will start.
c) Other depots to be established in the support and assembly trenches. These latter should be close to the junction of these trenches with the communication trenches.
d) A central brigade depot to be established still further in the rear from which these advanced depots can be replenished."

The locally controlled front line bomb dumps were small, so as to cut down the risks of explosion. Ideally they were dry, bomb proof, and placed at frequent intervals in the bombing trench and near the heads of the communication trenches. The best pattern of dump was thought to be a T-shaped trench off the main trench. Later plans would be issued showing how to construct bomb recesses and bomb stores, the main materials being planking, roofing felt, and corrugated iron. If possible bombs were to be stored in tin lined boxes. In addition to these supplies it was intended that men sent up the line in support of an attack should carry extra grenades, and that German grenades were also to be used when captured.

Grenades were supposed to have their detonators inserted in the area of Brigade Headquarters, ideally in sheltered conditions, with only one or two men exposed to the potential danger at a time. In practice things did not always work out this way, and the job was sometimes done close to the front, or in the open, with the troops nearby. On one occasion Guy Chapman recollected seeing some 300 men sitting in an open square doing this dangerous work:

"Suddenly there was an explosion and a shout. Pieces of iron hammered and slapped against walls. Three of the 60th [King's Royal Rifles] were lying in a bloody group: one holding his stomach and screaming. Stretcher bearers ran up and carried the wounded away. The detonating went on without pause."

The year 1916 would see further changes. Most obviously the men who threw grenades were no longer to be known as "Grenadiers," but "Bombers," the change at least in part due to a protest by the 1st Foot Guards, who were jealous of their historic regimental title "Grenadier." The semantic change was, however, the least important, because it was 1916 which would see the beginnings of a new flexibility, and determination to integrate different sorts of weapon. Above all it was now demanded that bombers, though

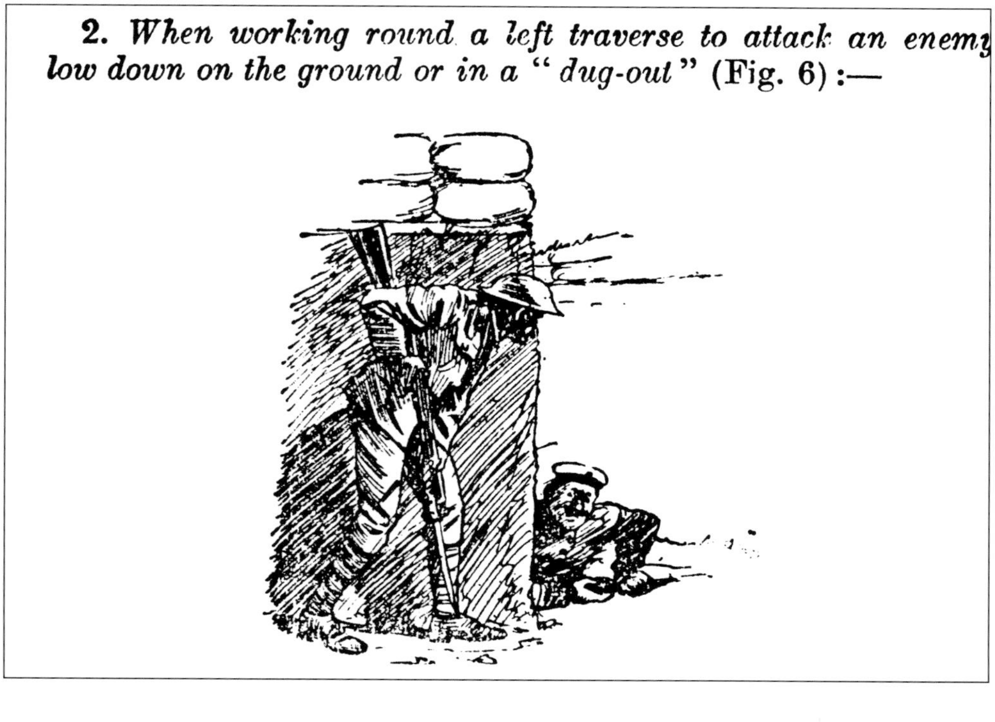

2. *When working round a left traverse to attack an enemy low down on the ground or in a "dug-out"* (Fig. 6) :—

Right: **Bombing squad tactics by the book showing how bayonet men should round a corner into a dugout.**

Far right: **A Royal Engineer prepares to fire a No 36 Mills bomb from the cup discharger.**

specialists, should not loose their facility with the rifle. The object of the Bombing Party was proclaimed to be "rapid and continuous advance," with grenade throwing "only to be resorted to when the bayonet men are held up." Moreover it was optimistically asserted that the action was never to degenerate into a bomb duel.

Unsurprisingly, this was not universally achieved, and a great deal of stress was placed on the bayonet men, who were specially selected for "pluck and alertness" and their ability to take "snap shots." Of the two throwers in the party, one was to take responsibility for the short throws, and the better of the two, the long throws. Dugouts were to be dealt with without delay – two grenades thrown down, and the bayonet men to follow after the explosions, guided by an electric torch. In raids the throwers were instructed to go armed with revolvers, bayonets, stabbing knives, or other hand-to-hand weapons such as axes or knobkerries. In the general assault all men were to carry rifle and bayonet, the bombers being taught to throw with the rifle slung over the left shoulder. Steel helmets were to be worn by all taking part.

A classic account of a 1916 bomb fight on the Somme was recorded by Lieutenant Charles Carrington of the Royal Warwickshire Regiment. First he described how he heard distant voices; then a moment of terror, when the mind was full of, "Prussian Guardsmen, burly and brutal, and bursting bombs, and hand to hand struggles with cold steel." Then came the bombs in reality, explosions filling the air with whining fragments, bay by bay, and closer with each moment. Carrington risked a look over the top:

"Thirty or forty yards away I saw a hand and a grey sleeve come up out of the trench and throw a cylinder on the end of a wooden rod. It turned over and over in the air, and seemed to take hours to approach. It fell just at the foot of the traverse where we stood, and burst with a shattering shock… Sergeant Adams pulled a bomb out of his pocket and threw it. I did the same and immediately felt better. A young Lance Corporal, Houghton, did the same. The next German bomb fell short. Then someone threw without remembering to pull the pin, and in a moment the bomb was caught up and thrown back at us by the enemy… I snapped off my revolver once or twice at glimpses of the enemy… I was watching them over the traverse, when I had the impression that someone was throwing stones. Suddenly I saw lying in the middle of the trench a small black object, about the size and shape of a large duck's egg. There was a red band round it and a tube fixed in one end of it… It was lying less than a yard from my foot; I was right in a corner of the trench. What was I to do? In an instant of time I thought: had I the nerve to pick it up and throw it away? Should I step over it and run? or stay where I was? There was no room to lie down. But too late. The bomb burst with a roar at my feet. My eyes and nose were full of dust and pungent fumes. Not knowing if I was wounded or not, I found myself stumbling down the trench with a group of groaning men."

By this time a separate sub-unit organization had come into use for rifle grenadiers, of an NCO and eight men. Although 1916 still saw the use of various rodded Hale grenades, a version of the Mills with a detachable rod, known as the No 23, was coming into use. This had the advantage that the same basic bomb could be used in both tactical roles. The training of the new group included firing, with and without the aid of sights and wooden stands, ranging and observation of fire, preparation of emplacements and firing at vertical targets.

Tactically the rifle bombers were intended to achieve ease of movement and concentration of fire, their prime targets being enemy machine-gun posts. Alternatively they could distribute their fire along a given section of

trench, thereby catching the enemy unable to run or otherwise avoid the barrage. It was of increasing importance that the rifle bombers integrate their efforts with other weapons, timely intervention in grenade fights allowing the rifle grenadiers to outdistance hand throwers. Equally rifle bombers could add weight to attacks on strong points and machine guns, a "sudden and rapid barrage" knocking out or suppressing the enemy's resistance allowing riflemen to advance.

Continued improvements were made to grenade training during 1916, and early 1917. Bombing was now organized on the premise that it was a general skill for all troops. A basic three day instruction course concentrating on the Mills bomb and throwing technique would give a reasonable proficiency, while an advanced course would deal with German bombs, catapults, and other devices. The training led to a two part test, the first consisting entirely of throwing dummy grenades, especially into, and out of, trenches. Part two of the test involved stripping different types of grenade, answering questions, and finally throwing ten live grenades.

Another much emphasized aspect of training was to link it with sport and competition, keeping the men entertained while imparting serious messages. The 7th (Territorial) Battalion, The Black Watch, for example held a sort of Highland Games at Bailleul-aux-Cornailles on May 10th, 1917. One of the events was firing rifle grenades from both standing and kneeling positions. More widespread was the officially encouraged game of "Bomb Ball." Two teams of 11 men threw a small weighted bag, similar to a grenade, from player to player, bomb throwing style. The object was to land it in the opponents' goal area; there was no running when carrying the ball, but passes were allowed in all directions, and procedures for throw ins, fouls and penalties all resembled soccer rules. This and other training "to prevent

Above left: **The No 34 Mk III grenade produced in late 1917; not as destructive as the Mills but lighter and handy for longer range.**

Left: **Contemporary color postcard showing James D. Pollock, Queen's Own Cameron Highlanders, who won the Victoria Cross for repeatedly climbing out of a trench under heavy machine-gun fire to bomb the enemy.**

Above right: **Body armor. Sketches showing how the early model Dayfield shield could be worn under the Service Dress Jacket.**

Right: **German *Stoss*, or shock troops, of Füsilier-Regiment *Anton von Hohenzollern*, Nr 40, pictured in early 1916. Their helmets are worn with covers but without spikes, bombers are festooned with stick grenades, and the men with their backs to the camera show how the equipment is worn in assault order with the shelter sheet wrapped around a mess tin.**

Left: **The 0.303-inch Vickers machine gun. Firing from a 250-round fabric belt at a cyclic rate of about 500 rounds per minute, the performance of the Vickers was similar to its German counterparts, but the British weapon was marginally lighter owing to its simpler tripod.**

Below: **Scourge of the Allied infantry: The German MG 08 machine gun on its sledge mount with crew. Note the metal belt boxes. (Courtesy Martin Pegler)**

Below right: **German gunners using the deadly MG 08 machine gun in an anti-aircraft role from a trench system in the Vosges.**

monotony" was published in October 1916 as *Games For Use with Physical Training Tables and Training in Bombing.*

It was during late 1917 and early 1918 that bombers were finally meshed with other weapons, to turn platoons and occasionally even sections into fully rounded miniature battle groups. The stem onto which the trainers grafted was the Bombing Party or Bombing Squad. By late 1916 a sniper and a couple of rifle bombers had been listed as optional extras for the Bombing Parties; through 1917 the practice would become ever more commonplace, especially where they were likely to encounter pill boxes, which, though exceptionally well protected, presented a point, rather than a linear, target. Not only were there specialist bombers within platoons, but it was usual to issue every man with two Mills bombs. What had begun as a black art practised by only a handful of men, with tiny supplies of bombs had become a staple of modern warfare. Where the "professional" BEF of August 1914 had had at best a few hundred bombs, the mass army of 1918, which would finally break the Hindenburg Line would have approximately 7.2 million available at any given time.

It is interesting that the "cult of the bomb" was a factor which helped to revive another idea of even more ancient ancestry, body armor. As early as 1914 there had been private initiatives, with some officers more or less shamefacedly donning "Heath Robinson" (Rube Goldberg) devices proffered to them by worried relatives, often under their Service Jackets. Yet, by the end of 1915, the idea had been picked up by the authorities. A call was made from General Headquarters to find out if a Bomber's Shield could be developed with the ability to afford all-round protection against splinters, which was neither so heavy, nor so cumbersome, as to render it impractical. Many tests were carried out in a special Fragmentation Hut erected for the purpose at Wembley, and diverse materials including various types of steel, silk, and vulcanized fiber came under scrutiny. Samples were made up and taken for trials. However, given the materials then in existence, the same types of problem kept recurring, as even if body armor could resist lower velocity fragments, it tended to fail with bullets. This was reported by Major Byron from France to the Trench Warfare Section,

"As regards a body shield it appears to be an open question whether a mobile shield of this nature is a practicable proposition, for to make it sufficiently thick to prevent penetration at 50 yards from the German rifle entails a prohibitive weight."

So it was that armor was slow to catch on officially, and was never worn by more than a very small proportion of British troops. Nevertheless, commercial patterns of armor achieved moderately good sales and photographs are seen of troops in Dayfield or Chemico body shields. Dayfield steel armors in khaki fabric covers also saw a limited official use. Experimental issues were made in June 1916, and the results must have been encouraging enough as, shortly afterwards, a request was made for 400 sets per division. These arrived in regular batches until mid-1917 when a lighter pattern was substituted.

Machine Guns

The machine gunner and his battalion officer went through as drastic a metamorphosis in the four years of war as did bombers and snipers. Again the forces for change were essentially threefold: Technological, environmental as encouraged by the trenches themselves, and tactical. Perhaps the most obvious feature to the casual observer was the introduction of the light machine gun. Although cavalry units used

Above: **Men of the new Machine Gun Corps, pictured at Shipley, Yorkshire. Note the crossed machine gun cap badges and the Vickers with both tripod and an auxiliary bipod under the barrel.**

Right: **Vickers machine gunners at rest on the battle field. Note the water and ammunition cans and other kit necessary for sustained fire.**

French type Hotchkiss light machine guns later in the war it was the Lewis which was issued *en masse* to the infantry and, it may reasonably be argued, this was the most successful new weapon used by the British Army in the Great War.

This was perhaps strange, because initially the Lewis gun was neither British, nor intended as a "light" model. As originally designed by Samuel Mclean the gun was a tripod-mounted medium or heavy weapon, which would be an alternative to the Maxim types then in use. It was only when the gun was further developed and promoted by fellow American Colonel Isaac Lewis, one time secretary of the US Ordnance Board, that the gun was recognized as an innovative departure. The mechanism of the Lewis was based on a turning bolt system, similar to that used in the Swiss Schmidt-Rubin rifle. Automatic fire was achieved by means of a gas-operated system, a piston being driven back after the first shot, withdrawing the striker, unlocking the bolt and rewinding a helical spring. About six air-cooled Lewis guns could be produced for the time and expense that it took to make one water-cooled Vickers. The Lewis was relatively handy, though it still weighed 26lb, but, lacking a water jacket, was not suited to sustained fire.

Despite its potential advantages the Lewis gun was not immediately adopted by the United States. British experts were soon aware of it as Colonel Lewis had visited Birmingham Small Arms to consult with the company on the making of barrels, and in November

Above: **A less than successful 1909 attempt at increased machine gun mobility. Invented by Lieutenant Strutts of the Derbyshire Yeomanry, and manufactured by Holmes and Company coach makers, Derby.**

Right: **The Lewis Light Machine Gun. The advent of the LMG was an important tactical breakthrough allowing machine gun support to accompany advancing troops at their own pace.**

1913 the new weapon was demonstrated at Bisley in the presence of British and foreign government representatives. The first tests were indeed remarkable, being successfully carried out from an aircraft against a white sheet on the ground. Terrestrial tests demonstrated the gun at 200 and 500 yards, and also showed that it could be fired inverted, or at any angle. As a result several governments placed trial orders. The handful deployed by the Belgians showed extremely promising results.

Although there were a very few weapons in service in late 1914 it would be mid-1915 before British production of the Lewis gun was sufficiently advanced for them to become a general issue, the objective then being to put four per battalion into the line. Total British production would eventually exceed 145,000 Lewis guns both for aircraft and ground use. Almost immediately there was a tactical impact. Lewis guns were lighter, easier to move up with an attack, or change position in defense.

They were not simply extra machine guns, but a new sub-category of weapon, the potential of which was as yet undistilled and ill formulated. Initial uncertainty about the role of the Lewis was reflected in the various ways it was at first categorized, being variously described as an "automatic rifle," a "light machine gun" or simply as a "machine gun." Lieutenant Basil Sanderson, Duke of Lancaster's Own Yeomanry, Divisional Machine Gun Officer with 23d Division in 1915, similarly noted how, in the early days, the question of tactical deployment was less than clear. On his own sector front 28th Field Company, Royal Engineers, resolved the issue by the development of a position which would accommodate either a Lewis or Vickers gun, "a very neat little emplacement with a platform and pivot mounting which slipped backwards into two different positions, according to whether you wanted to use a Lewis or Vickers on it."

The problem of how the different sorts of machine gun would be used was solved radically at the end of 1915. Vickers guns were removed from the infantry battalions and brigaded separately, while the provision of Lewis guns for immediate support within the unit was upped to 16, or one per platoon. The separate

Machine Gun Companies, containing the Vickers guns, were now lumped together as a separate entity, the Machine Gun Corps. This went one stage further than the normal German practice in which battalions maintained machine gun companies, and met with distinctly mixed reactions from the troops. As J.C. Dunn, medical officer serving with 2d Battalion, Royal Welch Fusiliers, had it: "This insane act was not reversed until the war was over, although from August 1918 onwards a Machine Gun Section was attached to each infantry battalion and acted under orders of its officer commanding." Even the official historian, Brigadier Sir James Edmonds believed that removing the Vickers and increasing the numbers of "not too reliable" Lewis guns weakened battalion firepower.

It is notable, however, that both these statements were retrospective, and that the period of the brigading of the heavier weapons coincided not only with the introduction of the Lewis, but ended effectively with the resumption of open warfare. The logic, no doubt, was that in trench warfare there was no significant disadvantage, and every tactical advantage, in removing the Vickers from the front line trenches to positions where they could provide longer range support, and indeed they were later organized to fire machine gun barrages at anything up to 3,000 yards. It is claimed that the first such Vickers barrage was delivered by 100th Machine Gun Company on the Somme, in August 1916:

"Captain Hutchinson requested that a rapid fire should be maintained continuously for twelve hours, to cover attack and consolidation. The gunners did that, and the gun proved its stamina value.

During the attack on the 24th, 250 rounds short of one million were fired by ten guns: at least four petrol [gasoline] tins of water besides all the water bottles in the Company and their urine tins from the neighbourhood, were emptied into the guns for cooling purposes; and a continuous party was employed carrying ammunition. Private Robertshaw and Artificer H. Bartlet between them maintained a belt filling machine in action without stopping for a single moment for twelve hours.

At the end of this time many of the NCOs and gunners were almost asleep on their feet from sheer exhaustion at their posts. The gun team of Sergeant Dean, D.C.M., fired just over 120,000 rounds. The attack was a brilliant success... Prisoners examined at Divisional and Corps HQ reported that the effect of the machine gun barrage was annihilating,

Left: **Mobile firepower, a Lewis gun team in the front line trench.**

and counter attacks which had attempted to retake the ground lost were broken up."

As *Tactical Summary of Machine Gun Operations* would explain in October 1917, detached machine guns could also be called down to provide what were known as "S.O.S. groups." These were patterns of fire on given areas, demanded by the infantry to deny ground to the enemy. The idea of neutralizing fire was further developed in *Lessons Learned From the Battle of Cambrai*, and the January 1918 manual *The Employment of Machine Guns*. This last described the role of the Vickers in set piece attacks, preventing reinforcement and supply, or making the enemy risk death by crossing swept avenues and crossroads. These things may not have been terribly economic, nor may they have given the infantry a good feeling of close support, but were tactically a good use of a heavy tripod mounted weapon with long range. Where this deployment fell apart was in mobile warfare, as was tacitly acknowledged by the replacement of machine gun sections in close support in August 1918.

The Lewis gun, which carried on within the infantry battalions throughout the war, similarly continued to evolve its own important tactical niche. They were controlled first at company level, later at platoon level. The Lewis gun section generally comprised eight men, led by a junior NCO, and apart from those actually allocated to fire the gun, and to carry it and its spares, three men were nominated ammunition carriers. According to *Notes on the Tactical Employment of Machine Guns and Lewis Guns*, issued in 1916, the Lewis gun was best used with the attack, or in the front line trenches.

On the defensive Lewis guns were to be used to economize on manpower, and to cover depressions and approaches which could not be seen from set machine gun emplacements. Although the Lewis would not now warrant a specially prepared emplacement itself, it was advisable to identify firing places, either depressions in the parapet, or loopholes. Methods varied, however, and Captain F.C. Hitchcock of 2d Battalion, The Leinsters, was one who preferred to dig Lewis posts in the form of small branches off the communication trenches, close behind the main fire trenches. In the offensive the Lewis was soon described as:

"*... particularly adapted for providing covering fire from the front during the first stage of an attack. Lewis gunners*

Left: **Reconstruction showing typical combat equipment worn by a Lewis gunner, *c.*1918. The large pack has been discarded, being replaced with a small pack. Wire cutters and entrenching tool are worn on the belt, with a harness for Lewis gun ammunition over the shoulders.**

Below and Right: **German machine gun teams. The most important German machine gun was the Maxim 08 7.92mm, water-cooled, belt-fed weapon that had a 600 round/minute rate of fire and a heavy sledge mount – seen to effect in the lower photograph.**

under cover of darkness, smoke, or artillery bombardment, may be able to creep out in front and establish themselves in shell holes, ditches, crops, long grass etc. where it will be difficult for them to be detected, and where they will be able to fire on the enemy machine gun emplacements, loopholes and parapets generally and to assist the infantry to advance."

It was not intended that the Lewis should be sent forward in the first wave against the enemy trench lines, but as rapidly as possible behind. In open warfare it might prove possible to have Lewis gunners with the leading line, where they would move, and appear to the enemy as, ordinary riflemen. In summary the Lewis was thought valuable for eight main purposes: To supplement fire, economize on infantry, reach places which could not be swept by heavier weapons, provide covering fire, help consolidate positions, reinforce lines, provide a mobile reserve of firepower, and lastly accompany "small enterprises" where a Vickers would prove too unwieldy.

Despite the havoc which German machine guns wrought upon British infantry on the Somme, the Lewis gun was one weapon with which the Germans felt totally outclassed in the summer of 1916. Writing on July 5th General von Stein remarked in his reports on the "large number of Lewis guns which were brought into action very quickly and skilfully in newly captured positions." He went on to say that, "It is very desirable that our infantry should be equipped with a large number of light machine guns of this description in order to increase the intensity of fire." The Germans would introduce their own standard light machine gun in the form of the MG 08/15, but it was never truly as "light" as the Lewis, and the Lewis gun remained a highly prized booty if it could be captured.

Methods of transport for the Lewis varied. In battle they were inevitably man-carried, usually across the shoulder, but in 1916 a sling was introduced for carriage when hot. On the march they could be drawn in small handcarts. Horse-drawn limbers available later in the war were capable of carrying four guns with spares, 22 magazine boxes, 176 filled magazines and 9,000 rounds of spare ammunition, a total weight of 1,949lb. Spare magazines were often carried by gun numbers in web bags, which came in several slightly varying patterns. A four-pouch carrier was introduced in July 1916, and appears in footage from the official film *Battle of the Somme*. Extracts from routine orders stated that indents for such carriers were to be submitted, one for every four magazines held. Formal *Instructions for Wearing the Equipment for Carrying Lewis Gun Magazines* were produced in 1917.

In December 1917 General Sir Ivor Maxse suggested that, in the light of increasing manpower problems, that the number of Lewis guns be scaled up. During 1918 this was deemed important enough that the total in each battalion was increased to 36, two per platoon, and four for anti-aircraft use with battalion

HQ. In the anti-aircraft role tubular metal tripods, very much like music stands, were issued. These supplemented rather than entirely replaced the multitude of improvised posts, upturned wheels and pintle mounts which preceded them.

From uncertain beginnings, Lewis gun training became an important part of battalion work. By the middle of the war each battalion had its own Lewis gun officer who was himself trained in army schools, one of the best known of which was at Le Touquet near Etaples. When out of the line the Lewis gun sections would be instructed by the Lewis gun officer, and experienced NCOs, with the aid of manuals, practical demonstrations and tests, as well as firing practice. The points to be taught to Lewis gunners were spelled out in detail in the document *Method of Instruction in the Lewis Gun*.

This divided the syllabus into sections, the first four dealing with general description, stripping, the mechanism, and points to be remembered during and after firing. The next three parts dealt with drill, cleaning and stoppages. This last was particularly critical, for despite its relative modernity, and great tactical benefits, the Lewis was quite prone to stoppages in the mud of the trenches. At least seven major categories of stoppage were identified, including several which could be cleared by "immediate action." There were, however, more serious problems which could relegate the weapon to single shots, or put it completely out of action for much longer. The final section of the

syllabus was a series of short practical tests. These included running five yards and bringing the gun into action in under ten seconds; a magazine change in three seconds; and swift magazine filling, as well as actual range shooting. Trained Lewis gunners were entitled to wear a badge. Initially this was identical to that worn by Vickers machine gunners, with MG in cloth or brass, but Army Order 80 of 1917 confirmed what already seems to have been happening, and MG was supplanted by LG for first class Lewis gunners.

In 1917 the Lewis gun would be integrated as the light support weapon of the platoon. Co-operation had already been encouraged between weapons with Lewis guns covering bombing parties, or flanking pill boxes, but now organization and theory were altered to match.

How this was to be achieved had been spelt out in *Instructions For the Training of Platoons for Offensive Action*, in February 1917, although it was some time before its tenets would become general practice. According to this document the guiding principle was that the platoon was a combination of all arms. The model platoon organization would be of four sections, each of about nine men, led by an HQ of an officer and four enlisted men. Each section was a specialist section in the sense that one was a bomber section; one the Lewis gun section; one a rifleman section, but with an emphasis on scouting and sniping; and the last a rifle bomber section. The new organization was to be suitable for both trench-to-trench attack, and for open warfare, so that, although the preparations and initial dispositions might vary, the actual tactics were often identical.

The Lewis guns, like the bomber and sniper, were no longer locked into attacking waves, but were often employed in "blob" or "worm" formations. These could themselves be arranged in a rough line, or deeper bodies. A gun section in the worm formation would "dribble up" to the attack; scouts would proceed the Lewis guns in bounds, on the direction of the section leader, and the remaining members of the team would follow up and flank the Lewis gunners. Fire was best directed, not straight to the front, but to the sides, assisting the advance of neighboring troops. Just such a formation was adopted by 2d Battalion, Royal Welch Fusiliers, near Hamel in July 1918:

"The companies carrying out the main attack were C, B, and D, to whom Engineers for demolitions were attached. They were to advance in line, each having two assault platoons on a front of 175 yards... Assault platoons consisted of a bombing section in each flank and a Lewis gun section between, the sections to advance in 'worm' formation. The C.O. allowed companies a large latitude in making their dispositions but he supervised everything."

Similar methods were employed by the Australian Corps in the same sector, using their Lewis guns in a truly "light" manner;

"In cases where a tank was not immediately available to clear up a hostile nest, one of the guns of the L.G. section, carried from a sling and fired from the hip, gave sufficient cover for the remaining gun to come into action deliberately. In conjunction with rifle grenades, fired by a proportion of

Left: **Going "over the top," Somme, July 1916. Tactical sophistication would only be evolved through painful learning.**

each of the rifle sections, sufficient assistance was provided for the infantry to overcome the local opposition."

So it was that rifle, grenade, and machine gun came to form a new tactical synthesis, totally alien to the precepts of 1914. Indeed it is arguable that infantry tactics, as exemplified by publications such as *The Division in Attack*, or *Training and Employment of Platoons*, of 1918, had moved faster and further than at any time in previous history. The stupid rifle-bearing automata of former days were now well on their way to becoming true specialists, educated rather than simply drilled in their trade.

In recent years debate has often revolved around the issues of whether German tactics were more advanced than British; whether the British Army ever actually fielded Storm Troops; and whether one system was essentially "better" than the other. Though interesting, these are questions which obscure rather than highlight the truth. The answers which have been given also tend to oversimplify the complexities. Neither the British nor the German systems were ever static, and to a large extent every combatant army learned both by experience, and by copying what were perceived to be the best tactics and equipment fielded by other nations.

Both the British and the Germans were organising bomb parties in 1915, though arguably the Germans were first off the mark. By the end of 1915 everyone was considering whether infantry attacks were better as limited objective advances or "big pushes." The Germans achieved early success with the flamethrower and the sniper; and while they maintained their superiority with flame weapons British and Empire snipers closed the gap. After a poor start British grenades arguably became better than the enemy counterpart, though their methods of use were similar and one copied from the other.

In 1916 the British achieved a clear lead with the Lewis gun, though as yet they had no answer to the enemy heavy machine gun emplacement. Nor any did they have a method which would grapple successfully with defense in depth. Similarly, during 1916, both sides were using school systems for specialist training, though arguably the formation of Storm Battalions as simultaneous shock troops and in-house trainers was a German innovation.

Both sides would learn much about the integration of weapons systems in 1917. By the final stage of the war methods were becoming increasingly similar, and training methods converging, though it was the British who had the greatest opportunity to make training universal and systematic. Yet still there were important differences: British tank superiority, German command initiative, which sometimes over reached itself, and chronic shortages of men and materials which would hamper both sides, but finally cripple the Germans.

CHAPTER THREE

THE GUNNERS

Artillery technology advanced rapidly in the years before 1914, and during the war the pace of change increased. Artillery became the major killer, causing more than 60% of all casualties. German statistics suggested that initially their losses were about 50% due to artillery, rising later to 85%. A British snapshot of the causes of wounds reported in January 1916 showed that 2,171 men out of 3,285 (or two thirds) of those recorded, had been injured by shells and trench mortar projectiles.

Among the prewar developments which made such carnage possible was the "Quick Firer"; a type of gun with hydrostatic recoil buffers and recuperators which effectively absorbed much of the shock of firing.

Left: **Desert trenches! Lewis gunner of the Black Watch in action from behind sandbags at Arsuf, June 8th, 1918.**

Below: **Workhorse of the Royal Artillery, the 18-pounder field gun.**

Although the barrel itself might whizz back with every bit as much venom as the guns of yesteryear the carriage would stay pointed at the target. Recoil was therefore much less of a problem. Combined with fixed charge ammunition, where the shell and its new cased nitro-cellulose propellant formed one handy unit, the result was much more speed and accuracy. Ease of operation was improved by rapidly moving screw-threaded breech blocks, such as were observed in action by Lieutenant John Stainforth of the Leinsters:

"It was a lovely picture; the gunners stripped and sweating, each crew working like a machine, the swing and smack of the breech block as clean and sweet as a kiss, and then a six foot stream of crimson from the muzzle, a thunderclap of sound, and away tore the shell over the hills to the Boche trenches 5000 yards away."

Typical of this category of QF gun were the Royal Field Artillery's 18-pounder, and the Royal Horse Artillery's slightly smaller, but very similar,

Above: **The German 77 mm field gun.**

Right: **German howitzer in action in Picardy in March 1918.**

13-pounder. Both were capable of a range just over 6,000 yards. On the outbreak of war these were organized in batteries of six guns, two horse or three foot batteries to an Artillery Brigade. Horses would remain the main motive power throughout the war, though tractors and trucks were seen in ever greater numbers. A single battery of 18-pounder guns in 1914 contained 21 wheeled vehicles, ranging from ammunition wagons to the water cart, and the battery bicycle. To keep them rolling needed 172 horses.

Shrapnel was the main form of shell in use, and though its invention dated back to Henry Shrapnel over a hundred years before, the latest versions were still remarkably effective against targets in the open. Large quantities of this type of round were used early in the war, so much so that the gunners would latter refer to the second day at Mons, as "Shrapnel Monday." The shrapnel shell worked by bursting above the target area, showering the ground with fast moving steel balls. Prewar training had stressed direct fire, firing at the enemy over open sights, rather than the more difficult arts of indirect and predicted fire. This was despite the fact that a 1911 staff conference had raised these points, and that two officers, Colonel John Du Cane and Lieutenant Colonel William Furse, had spoken eloquently in favor of the development of these techniques. *Field Artillery Training*, the main manual of 1914, was non-committal about the relative merits of fire directed by observers and telephones, but the high command and the infantry clearly expected the gunners close at hand. Thus it was, in the summer of 1914, that most fire was with shrapnel, and over open sights, as Lieutenant Hodgson of 122 Battery, Royal Field Artillery, recalled:

"The German Infantry appeared almost in front of us. There were hordes of them! They were in very close formation and they were coming forward, coming forward, closer and closer, nearer and nearer the guns. The order was shouted, "Gun fire!" That meant that all the guns had to fire at a speed of six rounds a minute. You can imagine the casualties they took. Men were falling like ninepins, but still they came on. But their tactics were that they moved forward a few paces, dropped to the ground, and then fired and came on

again. They were close enough to shoot at us with rifles, and we were firing back at point blank range. Of course this created sheer pandemonium."

Dramatic though these close actions were, and sometimes decisive, as L Battery, Royal Horse Artillery, would demonstrate at Néry, they were costly of men, guns, and ammunition. They would also become steadily less relevant as the war, the trenches, and technology progressed.

Ammunition expenditure was a growing problem. Scales of provision had been set at the time of the Boer War, and were not enough. Each 18-pounder gun had 1,000 rounds in the field. Just 176 of these were with the gun itself, and a further 202 were held at brigade and divisional level. The War Office establishments and the "modified Mowatt Reserve" formula allowed a grand total of 1,500 shells per gun. Monthly production of 18-pounder shells at first ran at a laughable 3,000 rounds per month; in short, if things were left as they were and no shells were actually fired, there would be enough ammunition to provide for an extra ten guns by Christmas 1914. Put another way if all the 624 18-pounder guns in Britain at the outbreak of war started rapid fire at once, all shells would be exhausted before the day was out. Not surprisingly the guns were soon short of ammunition, a problem which precipitated a major "shell scandal," which was to have political as well as military implications.

Less obviously, but equally importantly, it became apparent that the shrapnel shell was not best suited to the new warfare. Though shrapnel was good at killing, it

was only effective against targets in the open, and preferably crowded ones. Once men were in trenches they were relatively safe unless the shell burst directly overhead; and if they were in dugouts they were pretty well invulnerable since the shrapnel ball had minimal penetrative ability against the earth. What was needed to deal with field works was a shell which hit, or even entered the ground, before exploding. Again the Royal Artillery was ill served; a high explosive or HE shell was not tested for the 18-pounder until October 1914, and continuing production difficulties meant that it was still in short supply over a year later. So bad was the situation that only 9,100 filled high explosive shells had been completed for the 18-pounders by Christmas, and although production of shell bodies would pick up, there were still stocks of these unfilled shells in 1916.

The lack of high explosive shells for the lighter field guns would not have mattered so much had there been plenty of heavier guns and howitzers available. In fact the army had just 108 4.5-inch howitzers, and a mere 24 of the biggest gun type then available, the 60-pounder. Production of ammunition for these weapons was underway, but on a very small scale, and these guns also would suffer shell shortages.

In March 1915 Neuve Chapelle would be fought with guns totally inadequate to the task. Trench lines and dugouts were being assaulted with what were essentially field guns, designed to cope with targets in the open. At the heavier end of the scale there were a mere four of the new 9.2-inch howitzers known to the gunners as "Mother," and a single monster 15-inch howitzer, christened with commendable logic, "Grandmother." Arriving at the eleventh hour were eight 6-inch howitzers belonging to 59th and 81st Siege Batteries. These, almost by default, became some of the British Army's earliest exponents of predicted fire, using maps to determine their aim, as was noted by Bombardier W. Kemp:

"We could see only about two hundred yards on account of the trees, but we could see the spire of Neuve Chapelle church. The officer on this job was told to get a line on the church, hit it, and then register this as his line of fire and switch from it to other targets. He set up a plane table with a map of the area, put a pin on the church and another in the centre of the battery position. Then, with a director marked in a 180 degrees left and right, he took a zero line

Above: **10cm German guns in characteristic blotchy camouflage scheme.**

Left: **A long barrelled 6-inch gun in action. The men foreground left open the charge holders and set the fuses.**

Right: **A proud "Munitionette" with her product, Quedgley, Gloucestershire. The war brought new freedoms and earning power for such women, though the work could be dangerous, and was apt to turn their skin "canary" yellow through contact with explosives.**

Far right: **A new way of war: The *Materialschlacht*. The assembly room at Vanderveldt's plant in west London, showing British workers putting together No 5 Mills bombs to keep up with the "industrial campaign." (IWM 108429)**

from the pins and then gave individual angles to all the guns, which brought them all into parallel lines with the line of his director. One gun fired and hit the church, and the others took parallel lines to it… "

Another even more experimental method, attempted at this time, was the signalling of fire results back to batteries by Morse code, from radios mounted in aircraft. The outcome of tests was promising but it would be a long time before such a thing could become general. The progress made was published in *Co operation of Aeroplanes and Artillery When Using Wireless*, in July 1915.

If the types of artillery and their direction were not the best, at least concentrations of hardware were achieved. In all something in excess of 340 guns were gathered for Neuve Chapelle, one for every six yards of the front. Where finesse, super-heavy guns, and the shells for a lengthy fight were lacking, the mass of pieces concentrated in a small area, and the relative brevity of the bombardment did help to bring about at least localized success. As Lance Corporal W.L. Andrews would later record:

"The bombardment started like all the furies of hell. The noise almost split our wits. The shells from the field guns were whizzing right over our heads and we got more and more excited. We couldn't hear ourselves speak. Now we could make out the German trenches. They were like long clouds of smoke and dust, flashing with shell bursts, and we could see enormous masses of trench material and even bodies thrown up above the smoke clouds. We thought the bombardment was winning the war before our eyes… "

The euphoria was short-lived. The breakthrough never came, and the artillery would cease fire through lack of shells, their allotment having been barely enough for three days.

Eventually, by May 1915, ordering and planning of munitions production would be taken out of the hands of the War Office, and placed in the charge of the newly created Ministry of Munitions answerable to Lloyd George. The new minister rapidly selected "business men" to run his department, but the fruits of

reorganization were not overnight, as even Lloyd George admitted in his memoirs: "I cannot claim that my first choices were always the best," he stated ruefully. By July 1915 he had, however, come to some conclusions as to what had gone wrong, and who it was that he blamed for the state of affairs. As he reported to the House of Commons:

"1) The War Office had not made any thorough survey of its needs on the assumption that the British Army was to develop into a gigantic force of at least 70 divisions, and that its task would necessitate breaking through formidable double and often multiple lines of entrenchments, defended by masses of artillery, heavy as well as light, and thousands of machine guns, and that this operation would be a prolonged one. It had not, for instance,

come to any final decision as to the number and calibre of the guns which would be required for this purpose. The quantity and type of guns ordered were both obviously inadequate to the undertaking in front of it. Nor had it calculated the number of machine guns required for so great a force.

2) Until the number and especially the size of guns had been decided no one could be expected to compute the numbers and sizes of the shells required… "

This was all true; but it had never been up to the bureaucrats of the War Office to decide who Britain was to fight, or under what circumstances. Their job had been to create the best force possible, with the funds available, to meet the widest range of possible scenarios. Where Lloyd George's criticism was perhaps more telling was when he pointed out the War Office's failure in placing contracts widely with outside firms. In the meantime, for most if not all of 1915, and through the Battle of Loos, the army would have to struggle on, under instruction to co-operate in whatever scheme the French saw fit, with its most effective killing arm hamstrung.

It was not until 1916 that the artillery was equipped in anything like the manner required. By that time industry would be harnessed in a way hitherto unimagined, and shell production figures would be astronomic. In that year 52,943,513 filled shells would be provided, almost exactly ten times the production of 1914. In 1917 the figure would rise to 87,668,053, dropping back slightly to 69,809,834 in the last year of the war. At the Third Battle of Ypres the preliminary bombardment would cost £22,000,000 in shells alone. As with the shells so with the guns. At the start of the war the BEF had 378 guns in France. By the beginning of 1915 this had swelled to 749; a year later it would be 2,113, and by the end of the war 4,169. Howitzers showed an even more dramatic rise; from 108 in 1914, the number burgeoned to 2,985 at the end of the war.

Several entirely new categories of medium and heavy howitzer, like the 12-inch, joined the inventories. The use of these super-heavy guns could be dramatic in the extreme, and left an impression at the point of firing as well as the point of impact. The 12-inch rail gun firing from Dernancourt at the Germans near Bapaume on the Somme, for example, was witnessed to rip the tiles off nearby roofs, flatten the tents of an encampment, and cause horses to bolt in panic. A man in the 9th (Service) Battalion, The Princess of Wales' Own Yorkshire Regiment, declared that each discharge made his camp bed jump six inches. Yet the rate of fire of such a weapon was pretty slow, and the detachment amounted to about 70 men.

Lieutenant Sanderson of the Duke of Lancaster's Own Yeomanry watched the gunners labor at their task:

"Two men opened the breech, one checked the dial [sight], while two more operated the elevating gear. Then a trolley was wheeled along from the next coach to the breech of the gun with a shell on it. That took eight men. Eight more came along with a long ramrod & pushed the shell into the breech. The trolley rolled back again, and pushed by four fresh men came back with three colossal slabs of cordite. This was pushed in in its turn & the breech closed, when all the other various men assisted in laying the gun. I was standing on one of the coaches in the rear, with my fingers in my ears, & even then when the thing did fire, the shock I got nearly toppled me off."

"WHERE DID YOU GET THAT HAT?"

Below left: **"Where did you get that hat," from the Christmas 1915 edition of *The Dump* trench magazine of 23d Division. French and British soldiers admire each other's new "tin hats." the British Brodie helmet, which appeared in late 1915, was designed specifically as a defense against shrapnel.**

Right: **Increasing mobility: The FWD three ton truck. In this case of this particular type of vehicle cabs and chassis were imported from the Four Wheel Drive Company of Clintonville, Wisconsin, and fitted with bodies in Britain. FWD trucks were particularly useful as artillery tractors and supply carriers.**

Increases in weight and numbers were but part of the story, for improvements were not just quantitative but qualitative. Perhaps most obviously the use of Forward Observers, or FOs, became the norm rather than the exception. These brave souls, operating from the trenches or forward observation posts, usually communicated with their batteries by means of field telephone. They habitually received the thanks of nobody. The enemy targeted forward observers as a particularly threatening menace. If they failed in their duty the infantry and the command would criticize them for lack of support; if they succeeded they would be held to blame for the enemy response. The batteries were now located some way back, avoiding, it was hoped, the worst of the counter-battery fire, but able to range well over the "deep battlefield." Just such a set up was described by Lieutenant P.J. Campbell, Royal Field Artillery, who served with a battery of 18-pounders in France in the last two years of the war.

"When the battery was in action it was always split into two parts. There was the gun line, where the guns were, usually about two miles behind the infantry in our front line; and there were wagon lines, where the limbers for moving the guns were kept, and the ammunition wagons, and all the horses, two or three miles back, out of the range of most enemy guns.

The battery commander, generally a major, lived in the gun line; his second in command, a captain, at the wagon lines. The four or five subalterns were sometimes in one place sometimes in another. They took it in turns to come down to the wagon line for a rest. So did the gunners. The drivers were

always at the wagon lines with their horses, though they took ammunition up to the guns and drove up to the gun limbers before a move. Every gunner and every driver had his special position and responsibility in the battery, changeable only when casualties made change necessary.

The signallers, an elite in the battery, lived in the gun line. On them depended all our communications. The other specialists, shoeing smiths and veterinary sergeant, battery clerk, stores men lived at the wagon lines. The officers' servants went where their officers were. There were cooks in both places.

Our 18 pounder guns only had a range of six thousand yards. We could shell the enemy front line and fire at targets immediately behind it, but could not reach anything further back. More distant targets were engaged by the 60 pounders and six inch howitzers."

While the guns moved away from the front line, what they did tactically also changed as the war progressed. The Royal Artillery, already seen as a technical arm, would become the subject of innumerable pamphlets, range tables and manuals, which charted the path from simple point and shoot techniques, to a host of specialist tasks. Among the myriad booklets were titles like *Notes on the Destruction of Hostile Batteries by Artillery Fire*, of August 1915; *Notes on Artillery Observation from Kite Balloons*, November 1915; a lengthy series of *Artillery Notes* commencing in January 1916; *Co operation of Sound Ranging Sections and Observation Groups with Artillery*, November 1917; and *Effect of British Artillery Fire*, September 1918.

One of the first novel applications of artillery was the use of shrapnel for cutting wire. High explosive was rare in the first months of war, and in any case tended to have fuses which allowed the shell to bury itself before exploding. Shrapnel, on the other hand, could be set to burst above the ground, and the balls moved with speed enough to slice through wire. The theory did have considerable attractions, since, rather than throwing men against unbroken wire where they would be halted and cut down, there was a possibility that they could advance right into the enemy line. It certainly seemed a better option than hand-held wire cutters, and rifle attachments which exposed their users to near suicidal risk. Lieutenant K. Page of 40 Brigade, Royal Field Artillery, described the technique as it was employed during the great barrage leading up to the "Big Push" on the Somme on July 1st, 1916,

"I was in charge of a section of an 18 pounder battery and we were given the job of cutting lanes through the German wire. It wasn't an easy thing to do. You had to do it very slowly and very deliberately. You would go on plugging away at one short stretch of wire… and bearing in mind that there was wire all the way along the front, the tendency was for a gap to get cut here and then a gap got cut a little way along there and the infantry had obviously got to get through this wire, so they tended to get in the gaps and, if the Germans knew the gaps were there – after all they'd watched them being cut – they could line their machine guns up to cover them.

The experts, the 18 pounder battery commanders, were quite good at cutting wire, but it did need very careful laying because guns were rather inaccurate in those days. They had what was called a 'hundred percent zone'. That meant that, if you fired a hundred rounds from one gun, at, say, a range of three thousand yards and you then measured up very

Below left: **Firepower. Emplaced German 38cm heavy naval gun. Maximum range claimed, 46 miles, weight of shell 750kg.**

Right: **Typical impact of shell fire: Dead Germans near Blankaert.**

carefully the area in which all the shells had fallen, and you would then call that the hundred percent zone. But, although most of the hundred rounds – all laid the same way, remember – would be more or less gathered in the middle, quite a few odd ones would have exploded out towards the extremities of the zone. So, it wasn't easy to go on plugging one gun into the same hole every time…"

Cutting wire with shrapnel was no panacea. The key drawbacks were loss of surprise, a key ingredient to any attack, and waste of ammunition. Nevertheless wire cutting with field guns became a feature of most major operations in the middle part of the war. Another critical point was that the technique was by no means infallible and, if not carefully checked, could lead to disasters of the type recorded by the regimental historian of the Liverpool Scottish on the Somme at Guillemont in August 1916:

"Prompt at 4.20 the line moved forward, 'X' Company on the right, 'V' on the left with 'Z' and 'Y' respectively in close support. The enemy had already put down two counter-barrages, one on the support trenches and one in No-Mans-Land. The attackers soon ran into the latter barrage and also met terrific machine gun fire and were held up. Lieut. Colonel Davidson rallied the men and led a second charge himself in which he was wounded. This attack fared no better than the first. A third and a fourth time the battalion rallied and went on but with little better result. At two places a few men succeeded in entering the enemy's trenches but they were overwhelmed by numbers. At all other points those attackers who reached the German line found uncut wire and eventually were forced to return to their own trenches or to take shelter in shell holes in No-Man's-Land. So ended an attack which was doomed before it began."

The unit's losses were 17 officers out of 20, and 263 men out of about 600. A Victoria Cross, five Distinguished Conduct Medals, two Military Crosses, and eight Military Medals, were what the battalion had to show for these brave but totally futile efforts. This disaster, and many like it, were to form for many the enduring image of combat in the First World War. Perhaps fortunately these big attacks "over the top" were not the norm, but the exception, and emphatically not the stuff of trench warfare which was essentially a far more specialized and less costly undertaking.

The ability of artillery to cut wire improved immeasurably during the course of 1917, for not only did shells of all natures now become available in almost inexhaustible quantities, but a new fuse was introduced which made the use of high explosive against wire a much more attractive proposition. The 106 percussion fuse was sufficiently sensitive and fast acting to set off the high explosive at the instant of contact with the ground; the force of the blast therefore tended to spread out at ground level, not only dealing a devastating blow to human forms in the vicinity, but ripping wire from its pickets and blasting holes in entanglements. Cratering which slowed up the infantry was also reduced because the shells exploded at, rather than under, the level of the ground. "Preparation" time could now be cut to a minimum.

The artillery fire plan could, and often did, mean the difference between success and failure. On the Somme for example, though there was certainly a prolonged and impressive bombardment prior to the July 1st attack, there were good technical reasons why the guns could not and did not provide the sort of overwhelming advantage which Haig and Rawlinson, commander of Fourth Army, expected. One factor was the continued shortage of heavy artillery. While the French in their sector could manage a heavy howitzer for every 20 yards of the front, the British had one 6-inch howitzer for every 45 yards, and this was perhaps only half what the enemy had managed at Verdun. The German defenses in the relatively dry ground of the Somme were deep, and likely to be unscathed by light field guns. Counter-battery fire was similarly less effective than it would be later in the war. The British VIII Corps identified 55 enemy batteries before July 1st, but the Germans managed to keep another 11 undiscovered, by the simple expedient of keeping them quiet until really needed.

On one day that September 159 active enemy batteries were located; of these 70 were engaged with air–artillery co-operation, but only half of these were silenced, so attacking troops were thrown against sectors where the majority of the enemy guns remained in action. Details of the artillery fire plan could also be problematic, for while the bombardment was vast in scope it was little co-ordinated with the movement of the infantry. Over much of the front the attempt was

Above: ***Grannie*, a 15-inch howitzer with 1,400lb shells, seen in an Australian photograph taken near the Menin Road, October 1917.**

Above right: **From the sublime to the ridiculous. An officer of 444 Siege Battery, Royal Garrison Artillery, with friend at the breech of a 12-inch howitzer, near Arras July 1918. (IWM Q 6873)**

simply to batter down the enemy defenses, and follow this with an advance. Very often the enemy troops had time to gather their wits, emerge from their deep dugouts, and cut down the attackers with machine-gun fire.

In terms of technical innovation the creeping barrage was therefore one of the most important advances pioneered by artillery in the Great War. Again the idea was simple in the abstract, a wall of shell fire advancing in front of the troops which would either kill those emerging to confront the attackers, or simply deter them from appearing at all. Though theoretically possible much earlier, in practice the technique required things which were not routinely in place until at least mid-1916. An adequate numbers of guns and shells to stoke this storm of steel was obviously required, but reliable fire control, a co-ordinated fire plan, and such confidence in the arrangements that the infantry would feel safe enough from "shorts" to go close to the barrage were also prerequisites for complete success. Interestingly the development of the creeping barrage was encouraged by the infantry themselves, for there are recorded instances of troops under particularly aggressive junior leaders deliberately crawling out of their trenches in order to get maximum advantage from neutralizing shell fire even before such things were formalized.

The fire plans for creeping barrages could be incredibly complex, and some of the best contained subtle ruses to confuse the enemy. One particular favorite was to walk the barrage over the enemy position, and then walk it back again, a so called back barrage, catching any personnel who had popped up to deal with the expected attack. Very often the barrage was not one line of shells but several, making the beaten zone perhaps a mile wide. The line of shells nearest to

the attacking troops was likely to be poured down from 18-pounders and 4.5-inch howitzers; further back medium and heavy guns would concentrate their fire, perhaps to block likely avenues of counter-attack, or deal simultaneously with support lines. Such patterns might also include lines of shellfire preventing the enemy making sideways movements, or even box barrages which would trap and neutralize enemy concentrations or stubborn centers of resistance. A descant on this symphony of destruction would be provided by the shriller notes of machine-gun barrages, and localized coughs from the trench mortar sections.

One of the first, if not the first, creeping barrage was shot in front of the attacking troops at Montauban on July 1st, 1916. This was a straightforward curtain of shrapnel, but soon there were experiments with high explosive, gas and smoke. At Arras on April 9th, 1917 Third army mixed shrapnel and high explosive, half and half; at Cambrai in November 1917 Fourth Army used one third each of shrapnel, high explosive and smoke. The speed at which the barrage crept was critical to the success of the undertaking, and was often expressed in terms of the numbers of minutes taken to traverse 100 yards on the ground. Sometimes this would be as much as eight minutes, or expressed another way, less than half a mile per hour. While this may seem a snail's pace it had to take into account not only the infantry's ability to traverse ground, perhaps with occasional pill boxes surviving in it, but the artillery's ability to maintain a sufficiently thick curtain of fire. It became apparent that some of the best barrages were those that were most unexpected; where the enemy could be caught unawares lengthy preparation was superfluous.

General Plumer's assault at the Third Ypres in September 1917 was a case in point and marked a further step forward in the quality of artillery preparation. His plans appeared cautious in the extreme, but arguably cost fewer lives as a result. The capture of the Gheluvelt plateau would require no less than four separate steps, with a gap of about six days between each movement, and no movement to be more than about 1,500 yards. It accepted that only limited movement was possible, and sought to avoid excessive losses through over-ambitious advances or loss of communication. About 1,300 guns and howitzers were used, almost one for every three yards of front. The carpet of fire in front of the troops was 1,000 yards deep and made up of five separate elements. These were a shrapnel belt laid by about half the 18-pounders; a high explosive belt fired by the remainder of the 18-pounders and the 4.5-inch howitzers; a belt of fire laid to deal with enemy machine gun positions; a high explosive belt fired by 6-inch howitzers; and finally a zone beaten by the remaining heavy guns and howitzers. The heavy guns used half instantaneous and half delayed fuses on their shells, so dealing with both surface and underground targets. The methods were ponderous but, unlike on the Somme the previous year, the fire plan was generally successful. About 11 German counter-attacks were broken up by artillery fire before they could fall on the British infantry. The performance of the artillery was measurably better in mid-1917 than it had been in mid-1916, but still the result was not a real breakthrough, and when the weather deteriorated motion ceased.

Tactical innovations were mirrored by improvements to the artillery's eyes and ears. The significance of aerial reconnaissance, and the forward observer with his telephone have already been touched on, but the advances were greater in other areas. Surveying developed apace. The relatively inadequate maps of 1914 were first supplemented and then replaced by large scale maps of small areas, trench maps detailing field works, and photographs. Finally there would be a whole Field Survey Battalion for each army. Predicted fire was now made very much easier; eventually batteries could turn up at the last minute before an offensive, and take part with hopes of success.

Flash spotting was also a relatively simple science, but one which was only systematically developed after the trench lines had stabilized. The basis of the technique was that two or more observation posts would look for the tell-tale flashes from an enemy battery, and by simple trigonometry work out its position. The real difficulty was to make sure that the observers were actually spotting the same battery, rather than another fired by the enemy as a mask. One innovation which helped to make this work possible was Lieutenant Colonel H.H. Hemming's Flash Buzzer Board developed during 1916. At first the notion received little official encouragement, and the inventor was reduced to scouring the electrical shops of London for components.

Far left: **Dugouts, and a steel helmet on Kemmel Hill pulverized by giant shells, at least some of which have failed to explode.**

Left: **Veteran Lieutenant, Royal Artillery, France 1918. His jacket still has the old cuff rank insignia, and is heavily repaired around the ends of the sleeves. He wears the "wrist" watch which was more practical than the pocket watch in action, and a non regulation cravat.**

Sound ranging was a slightly more complex idea but somewhat similar in principle. Observers, or rather listeners, at different locations, can, by a calculation involving the speed of sound, work out how far away a gun is from their position. By drawing circles on a map it is possible, by determining their intersection, to discover the precise location of the battery. Basic experiments in the field were begun by Frenchmen Lucien Bull, and Charles Nordmann, and the Germans also made early advances. By 1917, British artillerymen using Tucker Microphones and Bull Recorders were doing their part in locating enemy battery sites. It was even possible to tell the approximate calibers of the guns firing by their pitch.

Effects of Shell Fire

The new concentrations of guns on both sides, the new types of shell, and the length of time to which the troops were subjected to them, brought effects which can scarcely have been imagined before the war. It was shelling which created the Dante's *Inferno* which is the picture of the trenches in the popular imagination; but shells not only blew men to fragments and decorated the trees with shreds of humanity, burying and reburying men to the stage where they had no known grave, they had new and profound physiological and psychological impacts.

Some of these effects were strange and subtle. Sometimes men inside a dugout died strangely peacefully without a mark upon them, victims of a nearby blast. When men were under heavy loads even a minor wound might cause them to topple forward and drown in mud. Sometimes men would see the first signs of psychological damage in themselves. Driver Harry Ainsworth of the Army Service Corps, who drove his truck across the Somme, would recall later the impossibility of missing dead men strewn on the road, and the weird appearance of British soldiers thrown up into the blasted boughs of trees. "You had to laugh," he said, "otherwise you would go mad."

"Shell shock," which is still difficult to define medically, could be either physical, or mental, or both in its origins. Fearing, seeing, even smelling death by shell fire could unhinge the strongest personalities; blasts could make a man "punch drunk" or gibber incomprehensibly. At first this did not receive classification as an injury, and was likely to be dismissed as malingering. Later the problem was defining how serious such a thing was, and distinguishing real damage from "funk."

One such "problem" confronted an officer of the 2/4th Battalion, The Royal Fusiliers, between Zonnebeke and Langemarck during the Third Battle of Ypres in September 1917.

"The stench was horrible, for the bodies were not corpses in the normal sense. With all the shell fire and bombardments they'd been continually disturbed, and the whole place was a mess of filth and slime and bones and decomposing bits of flesh. Everyone was on edge and as I crawled up to one shell hole I could hear a boy sobbing and crying for his mother… Depression, even panic can spread quite easily in a situation like that… I tried to reason with the boy, but the more I talked to him the more distraught he became… so I switched my tactics, called him a coward, threatened him with court

martial, and when that didn't work I simply pulled him towards me and slapped his face as hard as I could… "

Rifleman J.E. Maxwell of 11th (Service) Battalion, The Rifle Brigade, witnessed a more clear-cut case in a casualty clearing station, which had a near miss from bombs.

"They were very, very bad cases of shell shock, much worse than I was, and two of them in particular got up and ran amok in the ward with their hands over their heads, screaming and screaming and screaming. It was shocking as it was all in the dark… Then the doctor came, after the raid was over, and gave them an injection."

Shell shock could be bizarre, or even blackly comic under certain circumstances. Just such an instance was observed by Lieutenant A.G. May of the Machine Gun Corps during an attack near Wytschaete in 1917:

"At the same time the mines went off the artillery let loose, the heaviest group firing ever known. The noise was impossible and it is impossible for anyone who was not there to imagine what it was like. Shells were bursting overhead and for no known reason I thought they were some of our shorts and then I realised the Boche were putting up a barrage on our front line and no man's land… Not far in advance of our parapet I saw a couple of our lads who had gone completely goofy, perhaps from concussion. It was pitiful, one of them welcomed me like a long lost friend and asked me to give him his baby. I picked up a tin hat from the ground and gave it to him. He cradled the hat as if it were a child, smiling and laughing without a care in the world despite the fact that shells were exploding all around. I have no idea what happened to the poor chap…"

Artillery caused most casualties, but statistically speaking very few shells "had your name on them." British factories churned out 218,000,000 shells during the course of the war, of which nine tenths were used on

the Western Front. This suggests that it took more than 200 British shells to kill a German soldier, although the incidence of injury was higher. This sheer difficulty of blasting the enemy out of the trenches was itself a spur to greater invention and specialization.

Mortars and Other Weapons

Trench mortars and trench catapults may only have been artillery in the loosest sense of the term, and were often in practice manned by infantry personnel, but both were introduced to solve what were traditionally artillery problems. The basic difficulty was that from the gunner's point of view the trenches covered their potential targets all too well. The field gun had a relatively flat trajectory, and unless by some miracle the projectile happened to pitch right in the trench, or burst directly above, the occupants were relatively safe. High explosive increased the gunner's chance of doing harm to the trench garrison, but again, unless the shell hit very close to the trench, the effects would be limited. Howitzers threw large shells at high trajectories, which were more destructive, and, dropping from above, were more likely to smash trenches and dugouts. Even so howitzers tended to have minimum effective ranges, closer than which it was either not possible, or too dangerous, to fire. This was where the trench mortar and catapult came in, as they were designed specifically to drop their shells or bombs in, or close to, the front line.

For some time the Germans would have the best of it with their various types of trench mortar, or *Minnenwerfer*. Alert men could often spot the slow, heavy, *Minnenwerfer* projectiles falling through the air, and sometimes it was possible to take evasive action, but even if you were not blown to smithereens straight away, burial alive was an ever present danger, as was recorded by Captain Hitchcock of The Leinsters:

"B Company was again strafed by 'Toc Emmas' [Trench Mortars] and Lynch had several of his men buried by them in Boot Street. C.S.M. Hartley and I brought over half a dozen men with shovels to dig them out. Three men were dead from suffocation. We never expected to get them out alive, as we were digging for at least half an hour. Every few minutes our rescue work was interrupted by T.M.s which were directed at the section of the line where we were working. Several times we were within an ace of getting at the unfortunate victims, when a T.M. would be signalled. We would all dive for cover, and return to find the cursed 'Minnie' had landed slap in the same spot, and had obliterated the entrance of the dugout where we were working.

The rescue party were splendid, working away in their shirtsleeves, trusting to myself to warn them with my whistle... The unfortunate men who were buried had not a scratch on them when they were finally dug them out, but life was extinct."

The British Army had no trench mortars at the outbreak of war, but as early as October 1914 Sir John French was appealing for "some special form of artillery... which can be used with effect at close range in the trenches." Its desirable characteristics were later elaborated as mobility, great shell power, and accuracy at short ranges. A rash of ideas followed. Some were direct copies of the *Minnenwerfer*, others were new. One suggestion was a converted 75mm howitzer, but this was dismissed on the grounds of the danger caused to friendly troops by the tail piece from its bomb. The "fairly effective" weapon which was settled on, was the Vickers 1.57-inch Trench Howitzer, the design of which was approved in January 1915, with the first six actually reaching the front in March.

Above: **The British 2-inch trench mortar or toffee apple bomb thrower. Note the large yellow painted projectile and the periscope which allowed the crew to see out of the mortar pit.**

Left: **The West Spring Gun being prepared to shoot. The projector is held firmly down by sandbags while three men pull down on the cocking lever. (IWM 55588)**

Right: **One of the smallest mortars, the German *Granatenwerfer*, or "grenade thrower" which was widely issued during the latter part of the war.**

Far right: **Australians load the 9.45-inch trench mortar, or "flying pig."**

Alongside this pretty inadequate official response were mounted a range of emergency measures of greater or lesser success. One of these was a conversion of a 6-inch shell body, to produce a light "4-inch" mortar: Fourth Army took invention a stage further, converting iron water pipes to throw jam tin bombs, and accepting a gift of antique 19th century mortars from their French allies. Second Army cast its own brass 3.7-inch mortars. None of these were really efficient, and some were downright dangerous. One particular model, a smoothbore invented by Colonel Twining of the Indian Sappers and Miners, burst eight times out of 11 when the first issue was fired.

The many extraordinary trench catapults also belong largely to this early period between the fall of 1914, and the end of 1915. Many of these never got further than the trials stage, among them Lieutenant Frame's Centrifugal Bomb Thrower, and Monsieur Bonnafous' Rotary Apparatus for throwing grenades. Others like M.J. Dawson's Spring Arm Projector and R.T. Glascodine's Catapult, which were essentially modern versions of ancient siege machines, foundered before they could become officially adopted munitions. Two which did make it through trials to manufacture and official issue were the Leach Catapult, and the West Spring Gun.

The Leach, designed by C.P. Leach, and supplied by the London store of Gamages, comprised a wooden Y-shaped frame to which a sling was attached by means of India rubber springs. The sling was wound back by means of a crank, a grenade was loaded into the sling, lit, and then the catapult was shot by pressing on a trigger plate. Many soldiers, having a healthy respect for the lengths of twanging rubber, preferred to knock the plate with an entrenching tool handle or stick. First orders for Leach catapults were made in January 1915, and eventually there were 3,000 or more, the normal establishment to be maintained at 20 per division. Each weapon cost £6 17s 6d (£6.88/$35), and they were capable of throwing a 2lb missile about 200 yards.

The West Spring Gun was probably the most businesslike and permanent looking of the British trench catapults. Developed by Captain A. West in early 1915 it was a heavy metal construction on a wooden base, with a shooting arm and a cocking lever. The motive force came from 24 "bank springs" put under tension by cocking. Instructions for mounting stated that the spring gun on its base be planted firmly on level ground, and weighted down with sandbags. The setting lever was now pulled down "steadily and without jerking," the bomb was placed in the launching cup, and thrown by a "smart steady pressure" on the discharging lever. Maximum range, with a Mills bomb as the projectile, was 300 yards. Guy Chapman recalled seeing one in the trenches, and thought it was:

> "*... based on Caesar's catapults at the siege of Argusium. It was supposed to hurl a hand grenade with much force and accuracy into the enemy's lines. In practice it was more apt to shoot the missile straight up into the air to return to the marksman's head, supposing he still had one; for the machine was also calculated to decapitate the engineer if he was clumsy enough to stand in front of its whirling arm...* "

Any exaggeration was minimal, as the War Diary of 144th Infantry Brigade, 22 November 1915 confirms,

"Lt Schwalm, 6/Gloucs, Brigade Grenadier Officer was killed. While firing the West bomb thrower, his foot slipped and his head was hit by the arm of the machine, after the spring had been released. This was not the first accident which has occurred with this machine, a very cumbersome one from which the results obtained are no means commensurate with the dangers incurred by the user and the difficulty in manoeuvring it."

Both the Leach and the West Spring Gun were discontinued in early 1916; not it would appear because the role for which they had been devised had disappeared, but because better devices had been invented to fulfill the same function. Improved rifle grenades had shown themselves equally powerful, and much more portable, while trench mortars had developed considerably, and now showed an ability to deliver a much greater punch for a similar effort.

The mortars which would sound the death knell of the early improvizations, were the 2-inch Medium Mortar and the Stokes in 1916. The former was remarkable for its outlandish appearance and devastating power, while the latter was exceptionally light and handy, and is widely accepted as being the progenitor of the modern mortar. It is thought that the basic idea for the "2-inch" came from the Germans, who had published details of a design by Krupp as early as 1910. The secret of the weapon's lethality was the fact that though the launch tube was only two inches in diameter, the huge bomb, nicknamed the "toffee apple" by the troops, overhung the tube. The toffee apple (occasionally also known as the "tadpole") weighed 50lb, and was widely thought to be the largest munition that it was convenient to carry in a trench. Manufacture of this mortar was underway by mid-1915, but it was some months before it was arriving at the front in appreciable numbers.

The Stokes mortar, which experienced considerable teething troubles, was also put into production in 1915. Its beauty was its lightness and simplicity. Essentially it was little more than a tube with an internal diameter of three inches, into which was dropped the cylindrical bomb. The bomb was more complex than the launch tube, having at one end a firing cartridge which was activated by a striker at the bottom end of the tube. This launching charge threw the bomb to its target where it exploded on impact. At the front the advantages of the Stokes were immediately apparent; orders were soon placed for more than 3,000 weapons, and a supply of 176,000 rounds a week was demanded. There would be other models like the 6-inch Newton, and the hefty 9.45-inch mortar, but for much of the remainder of the war the 2-inch and the Stokes were standard issue. In the last few months of the war it would become apparent that the Stokes was one of the few "trench" mortars which was almost equally as useful in open warfare, as it could be easily carried in wagons or on a horse, and could be borne short distances by the crew themselves. By the end of the war every infantry brigade would have its own light trench mortar battery, while two heavier trench mortar batteries, manned by Royal Artillery personnel, were attached at divisional level.

Like the forward observers of the artillery, the trench mortar men were an ambiguous blessing to the ordinary soldier in the trenches, as was recorded by Private Henry Ogle of the 7th (Territorial) Battalion, Royal Warwickshires, near Hébuterne:

"It is a bad place for Minnenwerfer, but they do not worry me as sentry, for they are inaccurate, or so it seems, and may drop their 'Minnies' anywhere between the firing line and the village. On the whole the front line is safer than anywhere between. But the speciality of the fire bays where I am standing, from the reception point of view, is the rifle grenade… At present we don't seem to have the answer to it, though the trench mortar men pay us occasional visits, leaving before the enemy retaliates, invariably and heavily on the front line and supports. We do not like trench visitors of any kind and make no attempt to disguise our feelings towards the T.M.s. As a sergeant of ours remarked, looking with unfriendly eye to his opposite number who had just arrived with his men and guns, 'If you pop-gun wallahs came and stayed with us, you might be welcome, but you pops and goes. There's some as grouses about wot's all Take and no Give, but you lot it's all Give and no Take.'"

Right: **A Stokes mortar emplacement of the West Yorkshire Regiment. The small blue grenade over the corporal's stripes shows his qualification as a trained mortar crewman. (IWM 8461)**

Below: **German troops assemble a *Minenwerfer*. The heavier models were catastrophically destructive but difficult to move across the trenchscape.**

Artillery in all its forms had become more, rather than less, of a specialist weapon as the war progressed. Developments in artillery material between 1860 and 1918 first helped to establish the conditions under which trench warfare appeared, and then, through further and more rapid advances, helped to create an environment in which more fluid, but entirely new forms of warfare could emerge. It is also true that the guns and gunners of the Royal Artillery multiplied exponentially during the war.

Each British infantry division had been supported by 76 guns in 1914; this had fallen as a bald establishment figure to about 48 per division in 1918, but this was only a part of the picture. Not only had the number of divisions multiplied, but each division had a smaller infantry component. Moreover there were now trench mortars, a short range artillery in all but name, which added a further 18 tubes of ordnance to each division, and many more heavy guns which were held back to be deployed as reserves at vital points. Indeed if one counted the ratio of guns to men the proportion had risen from about six per thousand to thirteen. What was more the average weight of shell that these guns could fire had risen even more dramatically. Whereas the heaviest shell fired by British guns in 1914 had been 60lb, the heaviest shells fired later in the war would weigh 1,400lb each. Where stockpiles of shells in 1914 had counted hundreds and thousands, stockpiles in 1918 counted hundreds of thousands and millions. No serious battle was likely to be undertaken in 1917 or 1918 without expenditure of shells in seven figures. There was also a much higher chance that these shells would explode effectively, and, given the improvement in forward observation, survey, sound ranging and flash spotting, that they would go where and when they were wanted. Almost a third of all British troops in France and Flanders at the end of the war would be artillerymen.

By the beginning of the "hundred days" of open warfare in late 1918 the artillery was no longer ancillary to the battle plan, but a vital part of it. Each month it was thought that about 13% of the enemy's guns were actually being destroyed by the Royal Artillery, and many more were being temporarily distracted into artillery duels, or kept silent so they would not be knocked out. More than this the barrage was now not something which usually took place as a preparation, which would warn the enemy what was coming, but something which would happen simultaneously as a part of the attack. The shells fired in this comparatively short time span would number not dozens per gun but hundreds per gun. During General Rawlinson's assault at Amiens in August 1918 every 18-pounder was supplied with 600 rounds, every 4.5-inch howitzer had 500 rounds, and every 12-inch gun 200 rounds. Every other type of gun had 400 rounds. Where possible the infantry would contrive to make co-operation with the guns as simple as possible: Lieutenant General Sir John Monash's Australian Corps for example aligned themselves so that the barrage could fall on a straight line all along a 7,000-yard front. The start line tape for the infantry was placed just 200 yards short of where the barrage was falling; there was some risk of being hit by their own guns, but little opportunity for the Germans to emerge in front of the advancing troops.

So artillery would play a major part, not only in opening up the way for the infantry, but in the destruction which fell on the enemy, on what General Ludendorff would call the "Black Day" of the German Army, August 8th, 1918. Over 27,000 men were accounted for, many of them prisoners, and successful counter-battery fire would mean that the Germans had relatively little with which to reply. As a Fourth Army history put it the "enemy's batteries ... were deluged by a hurricane bombardment and neutralized to such an extent that the hostile artlllery retaliation was almost negligible." Artillery, recognized as a specialist arm before the war, had undergone a remarkable technological and tactical evolution.

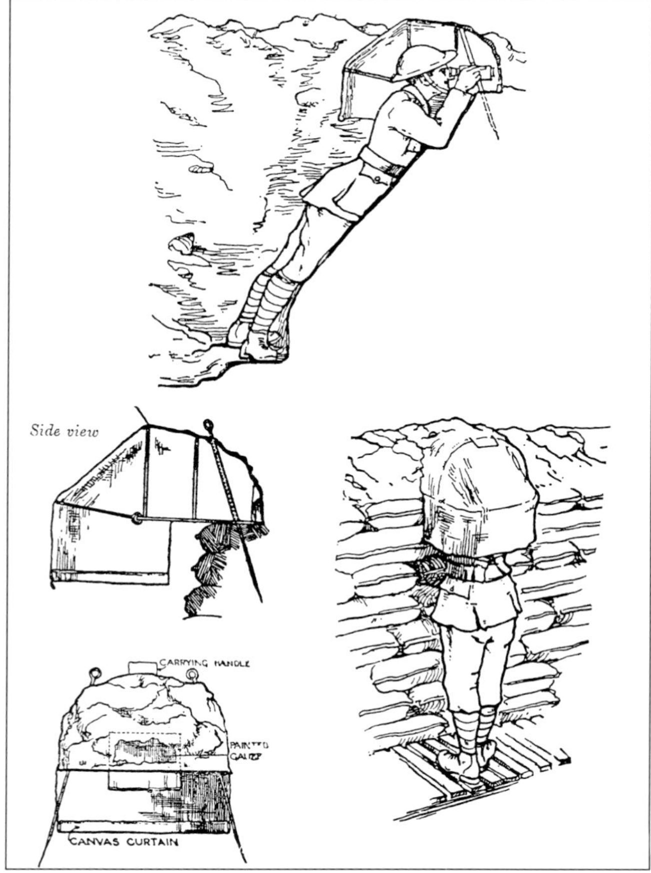

CHAPTER FOUR
GAS

While there have been many historical instances of the use of smoke, fire, and noxious stinks, to dislodge enemies from defensive positions, the Great War was the first war in which chemical compounds were scientifically identified, isolated, and used systematically for their toxic effects. That such an eventuality might arise had been recognized in the late 19th century, both in fiction, and in the arena of international relations. Appendix III of the Hague Convention of 1899 bound contracting parties to "abstain from the use of projectiles, the sole object of which is the diffusion of asphyxiating or deleterious gases." Even so this stricture was ambiguous and open to bending by those with the will so to do. It left open the possibility that gas might be combined with an explosive or projectile which had other effects; it was not clear that fumes which did not cause permanent damage were excluded; and it did not preclude releasing gas which was not in a "projectile."

Several countries adopted liberal interpretations of the agreement; others began experiments so as not to be left behind. New additions to the Hague Treaty of 1907, aimed against "poison" or "poisonous weapons," did little to stop the international community considering the options. The French soon possessed a small gas cartridge, while research in Britain in 1914 aimed at combining a "lachrymator," or tear gas, with a high explosive shell. The results of this were unimpressive and the program was suspended shortly after the outbreak of war, but work continued with hand thrown missiles, and a grenade was developed which released an irritant code-named SK, or "South Kensington." A mere 12 gallons of the stuff had been made by early 1915. By this time the Germans had

Right: **Men of 2d Battalion, Argyll and Sutherland Highlanders, wearing pads and anti-gas goggles, Bois Grenier sector, spring 1915. (IWM Q 48951)**

Left: **Portable camouflage. A hide made of canvas and gauze to conceal the observer.**

already used chemical weapons on the battlefield, though the first experiments on the Western Front in October 1914, and on the Eastern Front in January 1915, proved so unsuccessful that the enemy never realized that gas was being used.

Yet by April 1915 Fritz Haber's gas unit, 35th Pioneer Regiment, was dug in near Langemarck, awaiting the right wind conditions to release chlorine gas from cylinders. Despite more than one security leak the British and French troops in the trenches opposite remained oblivious and unprepared. On the afternoon of April 22d the valves were opened and a greenish-yellow cloud began to drift over the allied line. First impact was felt by French colonial troops of the 45th (Algerian) Division who fled; 1st Canadian Division, less badly hit, stood the test better. Wesleyan army chaplain Owen Spenser Watkins, working at an asylum just behind the front, witnessed the results:

"Gun limbers passed at the gallop, fugitive Zouaves and Turcos clinging to them. In a few minutes the road in front of the asylum was chocked with… soldiers and panic stricken peasantry from the farms and villages around. The story they told we could not believe; we put it down to their terror stricken imaginings – a greenish-grey cloud had swept down upon them turning yellow as it travelled, blasting everything it touched, shrivelling up the vegetation… Then there staggered into our midst French soldiers, blinded, coughing, chests heaving, faces an ugly purple color… speechless with agony… The enemy, in violation of every law of war, of civilisation and of Christianity, had descended to the use of asphyxiating gases."

The attack which followed the gas achieved a limited advance, but it was almost as if the Germans themselves had not appreciated the opportunity which the chlorine presented. Within days there were more releases, and soon gas would become a common accompaniment to offensive action. The German first use was a propaganda gift to the allies, which they were not slow to appreciate. As Conan Doyle put it, the German's had "sold their souls as soldiers." The high profile Ypres attack also freed the hands of British scientists, and spurred Sir John French to ask for substances with which to retaliate, though sensibly he decided to wait until significant quantities of gas were available before acting. Colonel Jackson's Trench Warfare Section duly applied itself to the problem, and by June cylinders of liquid chlorine were under test at the Castner-Keller plant at Runcorn, Cheshire. In British use chlorine would be code-named Red Star. At about the same time a second type of gas, carbonyl chloride, or phosgene came under consideration. Investigation showed that this could be more dangerous than chlorine, because phosgene did not cause spasm of the glottis, which limited the amounts ingested. Phosgene would be variously known as Lancastrite, Collongite or CG; when mixed half and half with chlorine the result was dubbed White Star. It would enter service in early 1916. The question of gas in shells was reopened, and these, too, would be ready by early 1916.

Like tunnelling, and various other specialisms, offensive use of gas was to be the province of the Royal Engineers. Brigadier H.T. Thuillier, RE, was made Director of Gas Services but under him Major C.H.

Right: **Primitive anti-gas methods. Men of the 1st Cameronians (Scottish Rifles) using a Vermorel agricultural spray in an attempt to neutralize gas residue in the trenches, 1915.**

Far right: **A Vickers machine-gun team in action, c.1916. Both men have gas helmets, the gunner wears the special reinforced "cape" or yoke for carrying the gun over the shoulder.**

Foulkes was deputed to take charge of the Special Companies then in the process of formation. Foulkes had no training in gas or chemistry. Nevertheless there was much to recommend him for the task in hand, for he was energetic, thrived on danger, and felt that no weapon was inappropriate with which to bring the war to a successful conclusion. Many of the men under his charge were recruited on the strength of chemical knowledge, though some never got to use this skill. Yet their specialist function was acknowledged by the fact that the special companies had no private soldiers; all were recruited and paid as corporals. This would cause some hilarity, with jokes in other units circulating about the "comical chemical corporals," and some anomalies, as for example when one of their number was reduced to the ranks for a misdemeanor, and thus became the only private soldier in the unit. The first gas company was briefly numbered 250th, but within the year there would be four companies identified as 186, 187, 188 and 189. By 1916 these would expand to a whole brigade of five battalions. Gas troops were identified by an arm band of red, white and green vertical stripes, possibly a hint of humor on Foulkes' part since he had taken up his new command the day that Italy joined the war on the allied side.

Within a few weeks the new recruits, and men transferred from other units, were moved from the Royal Engineers' base at Chatham, Kent, to Helfaut, near St Omer, where training proceeded, and material was gathered for forthcoming offensive operations. The target was 6,500 gas cylinders and crews ready for the assault at Loos in September. Initial plans had 2,850 cylinders clearing the way for I Corps, and 2,250 as preparation for IV Corps. Lieutenant W. Campbell-Smith described the apparatus:

"We had these cylinders weighing about 1 cwt; then we had pipes which consisted in general of two pipes, one bent at right angles 4–5 feet long with a bend about a foot from one end and a shortened pipe also bent at right angles, one end

attached to the cylinder and the other to the pipe so that the length of pipe was sufficient to reach from the top of the cylinder to the top of the parapet. It was obvious to us that we should lose time in transferring the pipe from one cylinder to another and that it would be difficult."

With a view to obviating the pipe problem there were efforts to obtain flexible metal hoses, and Lieutenant W.H. Livens began to experiment with rubber tubes. Simply getting large numbers of heavy cylinders into close proximity with the enemy was a herculean task. The first leg by train and ship brought the awkward cargo from Runcorn to Boulogne; from here trains took them to Bethune. Here they were unloaded into General Service wagons, and just behind the lines manpower took over, two men carrying a cylinder slung between them. Vermorel sprays were kept on hand in the trenches, with the intention that they should be used to neutralize any premature leaks. During the discharge it was planned that, as the gas was exhausted, smoke candles would take over, convincing the enemy that the air was still dangerous, and concealing the movements of the British troops.

In the event nothing like the predicted numbers of gas cylinders were available on time, and instead of a prolonged discharge of gas an alternating pattern of smoke and gas had to be substituted. Lieutenant A.B White of 186 Special Company, Royal Engineers, described the assault:

"At first the gas drifted slowly toward the German lines (it was plainly visible owing to the rain) but at one or two bends of the trench the gas drifted into it. In these cases I had it turned off at once. At about 6.20 am the wind changed and quantities of the gas came back over our own parapet, so I ordered all gas to be turned off and only smoke candles to be used.

Punctually at 6.30 am one company of the King's advanced to the attack wearing smoke helmets. But there was a certain amount of confusion in the front trench owing to the presence of large quantities of gas. We experienced great difficulty in letting off the gas owing to faulty connections and broken copper pipes causing leaks. Nearly all my men suffered from the gas and four had to go to hospital. Three out of the five machine guns on my front were put out of action by the gas.

Very little could be seen of the German line owing to the fog of smoke and gas. Our infantry reached the enemy wire without a shot being fired, but they were mown down there by machine gun fire or overcome by the gas. One or two made their way back and reported that there were seven to ten rows of wire uncut, and that nobody reached the front German trench. A report also came that the enemy were not holding their front line, but were firing from their second line."

Thus it was that the new technology was rendered less than decisive, partly because of problems with equipment, but mainly because of the simple fact that with cylinder gas a gentle breeze had to be blowing from one's own line towards the enemy. Nevertheless it is remarkable that a gas warfare arm had been produced in just over four months, and that Sir John French and the High Command were prepared to put so much faith, perhaps too much faith, in its usefulness. Cloud gas would become something of a British speciality, so

Left: **An officer instructs men of the Irish Guards on the finer points of the anti-gas tube helmet, Somme area, September 1916. (IWM Q 4232)**

much so that during the Battle of the Somme 110 cloud attacks were delivered, far more than were managed by the enemy. A dispassionate account of gas clouds was given in the German manual *Instructions Regarding Gas Warfare* in July 1916:

"Much more effective than gas attacks by means of shells are those in which gas clouds are used. For this purpose the enemy chiefly uses the irritant gas chlorine, which is contained in the liquid form in iron bottles under a pressure of 6–10 atmospheres. Many thousands of these bottles are built into the front line trenches. Their presence can only be known with certainty by penetrating the enemy's trenches, since the cylinders are only burst by direct hits with shells and not by shell splinters. On opening the valves of the bottles the liquid chlorine is forced out under its own pressure and produces thereby a hissing sound, which on a still night can frequently be heard at a considerable distance... The white clouds produced by the gas streaming out from the separate bottles quickly unite to form an opaque, clinging cloud bank, which under the influence of dilution and warmth, gradually rises, without thereby becoming ineffective. In very dry air, i.e. during great heat or intense cold, there is no formation of an actual cloud. The chlorine gas is then recognizable even from long distances, by its greenish-yellow color... Chlorine is somewhat heavier than air and therefore remains

stationary or flows along (like water) in trenches, mine craters, dugouts and hollows in the ground. The gas cloud will frequently flow round slight eminences so as to form so-called 'islands' which remain free from chlorine… According to the number of cylinders used, the width of the gas cloud may amount to several kilometers and its depth to many hundred meters, so that the time it lasts over our trenches can amount to 10 minutes or more… The chlorine… acts vigorously on most materials, but particularly on metals."

While clouds of chlorine and phosgene entered the battlefield repertoire as frightful, if unreliable, weapons, the search continued for more effective gases and ways to deliver them. Experiments were carried out not only by the army but the navy, and several new compounds were introduced in late 1915 and 1916. Among these were Jellite and Vincennite, both of which included prussic acid, and were known by the initials JN and VN respectively. VN took some time to produce in viable quantities and so JBR a mix of Jellite and arsenous chloride was used as a stopgap.

None of these were greatly successful. Chloropicrin was also introduced in 1916, and this received the code PS, standing for Port Sunlight, where it had first been produced. Compounds of arsenic, known as TD, DA and DM were all produced in early 1918, and might perhaps have gained greater prominence had not the filling factory at Morecambe been burned down at a vital juncture.

Perhaps the best known, and most feared gas of the war was Mustard Gas. This blistering agent which attacked the whole surface area of the victim's body was first used by the Germans in July 1917. Its effects were often delayed, a factor which meant that the careless were likely to carry on exposing themselves to it. One of the first to encounter it in the field was A.S. Dolden, a cook with the London Scottish, who chanced upon ground that had been hit by mustard gas shells, and was unable to identify the odour as gas.

"The next day Robertson was worse, and had to be led to the aid post with a bandage around his eyes, for he could not bear the light on them. There was a continuous stream of water running from my eyes, and they were extremely inflamed and very sore. I was in chronic pain, as my head, throat, eyes and lungs ached unmercifully. In addition the mustard gas had burnt me severely in a certain delicate part of my anatomy that is not usually displayed in public! We had forty casualties from the gas, one of which was fatal."

The British quickly christened this plague HS, standing, it was said, for "Hun Stuff," and set out to replicate it. Technically mustard gas was dichloroethyl sulphide diluted with a solvent, and would prove more difficult to manufacture than the substances hitherto used, but chemists at St Andrews University soon came up with a method. Initially it was thought that only 15 tons a week would be adequate production, but before the end of the war the target had been upped to 200 tons, though in the event this figure was never actually attained. Mustard gas always remained hazardous to make; in six months the Avonmouth factory recorded 160 accidents, and 1,000 burns, with a total of 1,400 illnesses attributable to the gas, among a work force of just 1,100.

Gas shells were also perfected, and these offered several potential advantages over cylinder release. Perhaps most obviously it was possible to place the gas more accurately with an artillery barrage and, although wind was still an important factor, it was now possible to pick out the enemy's rear lines or troublesome gun

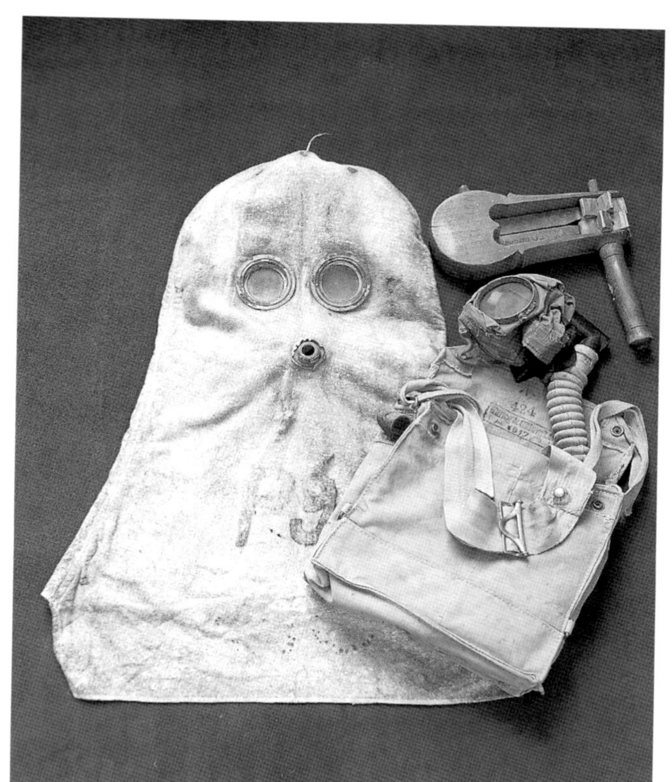

Far left: **The enemy. German bomber wearing the Model 1915 gas mask.**

Left: **Anti gas equipment, the grey flannel tube helmet, the Small Box Respirator" and a gas warning rattle.**

batteries and deluge them. Shell-delivered gas was also more likely to achieve surprise, bursting on a position rather than drifting onto the target. The results may not always have been lethal but putting a battery out of action temporarily, impeding its rate of fire, or discouraging the gathering of a counter-attack were all useful objectives. It was also true that shell-delivered gas was less likely to lead to friendly casualties, as it started to spread in the enemy's lines rather than your own. Against this had to be balanced the fact that cloud gas could be released in much bigger amounts and was likely to achieve much greater concentrations. British practitioners of chemical warfare seem to have remained generally biased against shell delivery; statistics from the end of 1916 suggested that while most other protagonists released three-quarters of their chemical munitions from artillery shells, only one-quarter of British gas was so fired. Nevertheless, the British gunners did use gas shells, and it was usually recommended that when they did so gas barrages should never be less than one hundred rounds, anything less having negligible efficacy.

Gas bombs were also made for the 4-inch Stokes mortar, but perhaps one of the most novel and cost effective methods of delivery was the Livens Projector which combined some of the better points of both artillery shell, and cylinder. W.H. Livens, now a captain in the Special Brigade, had begun work on a new method of throwing gas or flame in collaboration with his father, Mr F.H. Livens, in the summer of 1915. The nub of his idea was that a mass of simple inexpensive mortars could launch a barrage of large thin walled bombs, and thus quickly create a cloud of gas, or a barrage of flame, out ahead of friendly troops. Its advantages included the use of fewer personnel, and less complex emplacements for their use. Perhaps most significantly 50% of the weight of the Livens projectile could be payload in gas, whereas artillery shells could only carry 10%. First use of the new idea in the field, at La Boiselle, in July 1916, was not actually with gas projectiles but with burning oil canisters, and the projectors themselves were improvised by removing the tops from steel drums and planting them in the ground at 45 degree angles. A range of about 200 yards was achieved, and the results were encouraging enough for the improvised projectors to be used with gas later in the year at Thiepval and Beaumont Hamel.

Further tests continued, both at Porton Down, where an experimental installation had been established on a 3,000 acre site in January 1916, and at the front. These experiments produced a new gas projectile in which a Mills bomb was used as the bursting charge. Range data for the Projector was worked out by firing wooden projectiles with different charges and measuring the results. The weapons were now made in light, medium and heavy versions, the biggest of which could throw its bomb 1,300 yards. Despite production problems, and shortages of raw materials which led to changes in specification, the Livens saw widespread use and was adopted by the French and the Americans, and was about to be used by the Italians at the time of the cessation of hostilities. About 140,000 of these projectors were made in Britain between January 1917, and the end of 1918.

The Livens was never terribly accurate, but this did not matter since the idea was to saturate a general area quickly, or create a cloud over the enemy lines. Tactically they were planted in groups of about 25, and wired so that they could be set off by a remote electric firing system. Setting up was a dangerous job, but probably not as bad as dealing with gas cylinders, which were heavier, and more prone to leakage. Those on the receiving end of the Livens barrage were not slow to appreciate the finer points; according to one German report:

"The enemy has combined in this new process the advantages of gas clouds and gas shells. The density is equal to that of gas clouds, and the surprise effect of shell fire is also obtained… Our losses have been serious up to now, as he has succeeded, in the majority of cases, in surprising us, and masks have often been put on too late."

Remarkably the famous Projector was not W.H. Livens' only contribution to trench warfare, as his Z Company of the Special Brigade also helped to develop and deploy a number of flame weapons, though these were by no means as successful. The most extraordinary of these flame guns was the heavy or "large gallery" flame thrower. Weighing in at about two tons this had to be taken to pieces, delivered to the front line trenches, and reassembled on site. It shot a mixture of light and heavy Persian oil distillate, using deoxygenated compressed air as the propellant. It had a 100 yard range, the greatest of any flamethrower used in the war, and in its impressive but short ten second squirt it burned a ton of fuel. Terrifying and effective though it could be, its weight, bulk, cost, and difficulty in deployment was out of all proportion to the results, and so only five were ever delivered. They saw action on the Somme in July and August 1916, and at Dixmude in October 1917.

Livens also developed the somewhat optimistically termed Semi Portable flamethrower. This was rather like a large milk churn made of galvanised steel, and stood about 4 feet 6 inches high. It weighed about 150lb when filled with ten gallons of fuel oil and a similar volume of pressurised nitrogen. Its crew of two had the hair raising task of dragging the beast into position, attaching a tube, opening the valve, and then lighting the jet by means of a match. The pressure was

Far left: **Gas drill for Indian troops wearing the Small Box Respirator.**

Left: **A British Royal Engineer attaching a message to a bird using the Gas Proof Homing Pigeon Box, 1918. The soldier wears the Small Box Respirator, or SBR, which was introduced in August 1916, and continued in use with only minor improvements until after the war. The SBR, one of the most effective gas masks of the war, was also adopted by the US Army. (IWM Q 9288)**

such that the equipment had to be held down firmly, the crew hanging on "like grim death" when firing, lest the whole contrivance turn over or recoil on its users. These Semi Portables had a range of about 40 yards, and it appears that they were never brought close enough for effective use. They were discontinued entirely in the fall of 1916, leaving the Germans, who had managed to create a workable if somewhat unwieldy back pack model, with a distinct lead in the field.

Effects of Gas

Gas casualties are perhaps one of the most emotive subjects of the Great War, and whatever is said it is unlikely to expunge the enduring image captured by the poets. The most eloquent of these descriptions is probably that of Wilfred Owen in *Dulce et Decorum Est*:

"Gas! Gas! Quick, boys! – An ecstasy of fumbling,
Fitting the clumsy helmets just in time,
But someone still was yelling out and stumbling
And floundering like a man in fire or lime.
Dim through the misty panes and thick green light,
As under a green sea, I saw him drowning.
In all my dreams, before my helpless sight,
He plunges at me guttering choking drowning."

The victim is described, when dead, as having a hanging face "like a devil's sick of sin," with "blood gargling" from froth-corrupted lungs. However horrible the slow dissolving of one's lungs, neither the effects of chlorine nor the effects of phosgene, which could be delayed, were as bad as those of mustard gas. Mustard, as a blistering agent, slowly burned the skin off a victim's body. One fatal casualty, of the many, was described in the official medical history. First he turned brown over large parts of the body, with marked "superficial burning" of the face and scrotum. The gas also attacked the lungs, trachea and bronchi; and the liver became fatty and congested. His stomach had various hemorrhages, and his brain was wet and "congested." Another unpleasant aspect of Yellow Cross or mustard gas was highlighted by an American report which noted how, even in warm weather, the ground would remain polluted for two or three days after the impact of a gas shell. In very cold weather the danger would linger much longer, making digging in such ground hazardous even a month later. About 80% of British gas casualties from late 1917 to the end of the war would be the result of mustard gas.

Gas casualties suffered lingering malaises after the war, the lucky with slight coughs, the unlucky crippled by an inability to breathe properly, or blindness. A number of such unfortunates would survive into the 1990s. In the face of such awful symptoms it is difficult to remain objective, yet the remarkable thing about gas is not how many died, but how many survived. Of those

Right: **An Australian checks wind direction in the trenches near Armentières, May 1916. A very gentle breeze was crucial to the success of gas attacks, and for cylinder release this had to be in the direction of the enemy.**

Far right: **Loading a battery of Livens projectors ready for a gas shoot. The projector tubes have been dug into the ground, and are now being loaded with gas cylinders by gum-booted troops. The lieutenant in the breeches, left, has in his hands the electric firing gear. (IWM Q 14945)**

hit by bullets or shell fragments perhaps a third died, but gas was a different story. Of every hundred who became gas casualties only about three died, and a couple more were permanently invalided. Against this tiny proportion 93 of every hundred would return to duty, most of them within six weeks. Late in the war the Germans would become a little more susceptible to the effects of gas, perhaps because the men were in poor condition already, or because there were ever more new recruits taken by surprise by Livens bombs, but again the total proportion of dead never seems to have exceeded 6%. According to official British statistics just 307 men were killed by gas in 1915, and in no year would the tally reach more than 3,000. In short, out of more than 750,000 British deaths, fewer than 7,000, or 1%, were attributable to gas.

Yet the psychological impact of gas was out of all proportion to its lethality. As one veteran would put it:

"*With men trained to believe that a light sniff of gas might mean death, and with nerves highly strung by being shelled for long periods and with the presence of not a few who really had been gassed, it is no wonder that a gas alarm went beyond all bounds. It was remarked as a joke that if someone yelled 'Gas', everyone in France would put on a mask. At any rate, the alarm often spread over miles. A stray shell would fall near a group at night. The alarm would be given. Gas horns would be honked, empty brass shell-cases beaten, rifles emptied and the mad cry would be taken up. It sounded like the Chinese trying to chase off an eclipse. For miles around, soldiers woke up in the midst of a frightful pandemonium and put on their masks, only to hear a few minutes later the cry of 'All Safe'. Then they would take them off again amidst oaths and laughter. Two or three alarms a night were common. Gas shock was as frequent as shell shock.*"

One device for gas warning introduced midway through the war was the Strombos Horn. This was a megaphone-like piece of equipment, often permanently set up in the trenches, which would be activated when gas was detected. According to Frank McDonald of the Liverpool Scottish it sounded "just like a ship's siren a long way off at sea." Captain Christopher Stone of 22d (Service) Battalion, Royal Fusiliers, recalled how gas produced a "horrid" sensation even if ineffective, tending to leave him in a "rattled state," worrying much more than usual about details. A row of "gas gongs" was a prominent feature of a trench near his position, bearing the legend, "If the German gas you smell, beat these blinking things like hell." Some gases had a genuine mind-altering effect, and the psychotropic nature of the arsenical compounds was recognized in 1917. Depression and weariness which were chemically induced not only compounded the physical attack on lungs, eyes, and skin, but also opened the way to new accusations of malingering and cowardice. It was ever more difficult to distinguish between tangible ailments, phobias, panics, and a simple unwillingness to carry on.

Protective Measures

After a shaky start Britain would become one of the world leaders in defense against gas. The very first attempt at anti-gas protection was almost certainly a handkerchief held over the nose and mouth. Experiment discovered that slightly damping the cloth gave a modicum of protection, while using an alkaline solution, like urine, would, to some extent at least, neutralize chlorine. 1st Canadian Division placed dixies of water in the trenches, into which cotton bandoliers and handkerchiefs would be dipped and held to the face.

Major Matthews of the Winnipeg Rifles described dealing with gas with such rudimentary equipment:

"It is almost impossible for me to give a real idea of the terror and horror spread among us all by this filthy loathsome pestilence... many of course were absolutely overcome and collapsed to the ground, but the majority succeeded in manning the parapet... when the fumes were full on us breathing became most difficult, it was hard to resist the temptation to tear away the damp rags from our mouths in the struggle for air. The trench presented a weird spectacle, men were coughing, spitting, cursing and grovelling on the ground trying to be sick."

Slightly more permanent solutions were being sought behind the lines. Within days of the first attack some units were improvising their own primitive masks: 27th Division for example, arranged for lint strips with tapes to be made up by the nuns at Poperinghe convent. At Bethune, Army Workshops began the production of a pad soaked with lime water, based on the chloroform anesthetic inhaler, and within a week 60,000 would be made. Winston Churchill suggested a general issue of cotton wool pads, similar to those used during Admiralty smoke screen experiments. Despite the fact that these were of minimal use against chlorine the War Office launched an appeal to the public published in the *Daily Mail*. Over 30,000 of these devices were made in the first 36 hours, though breathing through them if they were wetted with a neutralizing solution was pretty well impossible. The uselessness of the early pads was one of the factors which prompted the invention of Barley's Respirator. Lieutenant Leslie Barley was a Territorial officer, serving with 1st Battalion, The Cameronians, who had taken an MSc in chemistry at Oxford. In early May 1915 senior officers set him up in a school science laboratory with the object of devising a

Right: **"Help us Win!" A poster by Fritz Erler in support of the German War Loans drive of 1917. Gas mask and grenades are at the ready.**

better gas protection. What he came up with was a pad of cotton waste inside a muslin bag, soaked in sodium hyposulphate and sodium carbonate. It would provide a measure of protection against chlorine, bromine and nitrous fumes. Barley demonstrated his respirator along with a crop sprayer which could be used to neutralize gas; the sprayer appears to have been a precursor of the later, widely used, Vermorel spray. About 80,000 of Barley's masks would be made. In the meantime the War Office had not been entirely inactive. On April 27th, 1915 a German prisoner was captured with a simple cotton waste type respirator secured by means of tabs; the chemist Herbert Baker was set to copying this. His achievement was an easy-to-produce replica, known from the material of its outer covering as the Black Veiling Respirator. Orders were placed in early May with the London chemical company Bell, Hills and Lucas. As a stopgap it was acceptable, but it would only work for a short period, and was prone to leaks.

All pad type masks required careful fitting if air was not to pass around the edges, and it was also true that most, if not all of them, required separate eye protection in the form of goggles. The obvious answer to these problems appeared to be a mask or smoke helmet which was pulled on over the whole head, and had integral goggles or a window for vision. Captain Cluny MacPherson of the Newfoundland Regiment was one of the prime movers in this area of research, and the idea of a full helmet was put forward even as the plan for the Black Veiling mask was nearing completion. It was quickly realized that the helmet was likely to be superior, but with time of the essence the pad masks were employed temporarily while the manufacturing details of the helmets were worked out.

Manufacture of the H or "hypo" helmet began in June 1915. First examples were of silver grey flannel dipped in a neutralizing compound, with a mica window. Although they were a massive improvement on what had gone before, the mica proved liable to cracks, and the flannel was in short supply. Later examples therefore made use of celluloid windows, and flannelette was substituted for the flannel. In all about 2,500,000 of these "hypo" helmets would be made. In the meantime research continued at the Millbank Royal Army Medical Corps center, using both human and animal guinea pigs. Often the work started with a rat in a box, the box being protected by material impregnated by the latest substance for testing. If the rat survived the testers moved on to a pig. If the pig survived a human volunteer would wear the new material.

During July 1915 a new model of helmet, the so called P or "phenate" helmet, was the fruit of this research. The main distinction was that the compound with which it was impregnated was sprayed rather than dipped, and had been altered so as to afford protection against a wider range of chemical agents, including phosgene and prussic acid. This new model mask was provided with a double skin of flannelette, round eye pieces, and a short tube to breathe out; 9,000,000 of them would be made. Finally, in early 1916, came the last helmet masks, the PH or "phenate-hexamine," and its variant the PHG. These later helmets featured improved chemical protection, and retained the breathing out tube and round eye pieces. The PH would become easily the most common of the helmet respirators with 14,000,000 produced, while 1,700,000 of the PHG were made. Captain F.C. Hitchcock of The Leinsters described the new gas helmets in his journal:

"*Got issued with a new gas mask, a bag made from Army flannel to put over the head. It had talc eye pieces and a rubber breathing tube for the mouth; the ends of the respirator were to be tucked under the jacket collar. They were all drenched with a solution of hypo, and were very sticky messy gadgets.*

"*We then marched the men off to watch a gas demonstration, a section of the Belgian defence trenches were covered over, and turned into a gas chamber. The officers put on their new masks and went into it. The gas was pumped out of a large cylinder. The gas was hostile looking stuff, and was a greenish yellowy vapour. When we came out of the chamber our buttons were blackened and our watches stopped, mine jibbed for ever afterwards, so I had to scrap it.*"

There would be just one more variation on the theme, but for animal rather than human use. These oddities were pigeon basket covers, pieces of material impregnated with anti-gas compound which wrapped around the boxes in which carrier pigeons were transported. Messengers in flight could of course expect no protection.

Useful though they were, the helmet pattern respirators never provided total protection, and would certainly fail with prolonged exposure to the more virulent gases. Among problems encountered were helmets which had rotted, soldiers who failed to breathe out through the tube, and irritation from the chemical impregnations. As early as the summer of 1915 it had been realized that the most certain method of protection would be an impermeable mask with a tube, attached to which would be a box of of solid absorbents. Charcoal was a useful absorbent, and various woods could be used for its production, but coconut husk was found to have some of the best qualities. In the United

States this would lead to the most bizarre wartime campaign of all, "Eat More Coconuts," sponsored by the Gas Defense authorities.

Even better than coconut charcoal was a soda-lime-manganate granule, the formula of which was communicated to Boots, the chemists of Nottingham, in December 1915. Boots would take these granules into production with the result that they commonly became known as the "Boots granule." The first mask to which they were commonly applied was the Large Box Respirator or "Harrison's Tower" a bulky piece of hardware, mainly issued to gas troops, artillerymen, and engineers in the spring of 1916. A more handy version, commonly known as the Small Box Respirator began to be issued in August 1916.

The Small Box Respirator or SBR consisted of a loosely fitting mask, made by pleating a flat piece of proofed material, which was impermeable for considerable periods to all gases, and thus provided protection for the eyes against irritants. It was provided with a mouth piece, nose clip and goggles, and gripped tightly across the forehead, down the sides of the cheeks and under the chin, being held in place with elastics and tape. One minor drawback of the mouthpiece arrangement was that a fair set of teeth was required to grip it, and it is recorded that at least a few of the toothless had to wait for dentures before proceeding overseas. Temple pockets were provided in the mask so that the goggles could be wiped by inserting the forefinger into the pocket. The nose-clip ensured that all inhalation and exhalation took place through the mouth piece. The mask was connected with the box by a metal angle tube and a rubber breathing tube. The valve for expiration was fitted to the angle tube, while the air to be breathed came in through a valve at the bottom of the box.

Familiarization procedures varied, but some units, even out of the line, adopted a practice whereby the mask was worn for a certain period each day. Such was the case with the 20th Hussars,

"In order to get all ranks quite familiar with their box respirators, these instruments of torture had to be worn for half an hour daily, whatever they were doing. Thus it might well happen that at Squadron Orderly Room a prisoner might quake in his boots and box respirator in front of a Squadron Leader who was endeavouring to 'tell him off' without dimming his eye pieces or shifting his nose clip. Witnesses would be in a similar state while telling 'the truth,

Right: **Sounding a gas alarm bell, Beaumont Hamel, December 1916. The mesh material along the trench is "expanded metal," sometimes known to engineers as XPM. (IWM Q 1717)**

Far right: **The full marching order equipment of a British infantryman, c.1916–18, complete with rifle, 1908 Pattern webbing, steel helmet, entrenching tool, bayonet, large pack and water bottle. The two bags either side are for left, the Small Box Respirator, and right, the old gas or tube helmet. (IWM Q 30217)** *the whole truth and nothing but the truth'. If the 'Sanitary man' made as an excuse, for letting the incinerator out, the statement that he could not see out of his respirator, this was merely taken to indicate that he required more practice."*

Anti-gas training became increasingly formalized with the progress of the war. In 1915 the few experts available had concentrated on impressing the troops with the importance of their masks and familiarization with gas, often by the simple expedient of travelling around with a cylinder and giving trainees a hopefully not incapacitating whiff of the stuff. By early 1916, gas, like so many other facets of the new warfare, was the subject of established schools. The experts and mask designers now trained a new breed of gas instructor, and these contrived to pass on the knowledge to as much of the soldiery as possible. Despite some failures and the rapid pace of development the results were generally good.

While the Small Box mask was pretty well universal there were minor changes made to it. The celluloid eye pieces were replaced by splinterless glass, and in early 1917 additions were made to cope with toxic smoke. This last was found to be best tackled by mechanical filtration, and so a pad of dry cotton wool in an "extension box" was issued for those masks already in use, or put into the existing boxes of new masks.

The Small Box Respirator was one of the best, if not the best, piece of protective equipment issued during the war, by any nation, and 13,500,000 were made. At the peak of production private contractors and the National Factories between them were fabricating up to 50,000 masks in a single day. In addition to new masks a total in excess of 2,000,000 were reclaimed and refurbished. Such protection came at a price, however, for they were relatively expensive to make, and each one had 105 components. Something like 30 factories were used for their manufacture, perhaps the most spectacular of which was created by taking over and refitting Tottenham Hotspur's soccer ground. Quality control would prove vital, and in one instance a batch of mask face pieces, numbering tens of thousands was rejected because of the inexact sewing. Whatever the drawbacks and discomforts of its use, the SBR would become pretty much the soldier's closest friend. During the retreat of March 1918 checks on stragglers discovered 6,000 who had dumped their weapons, yet only a mere 800 were without their respirators. Both America, and Italy, who discovered deficiencies in their own masks, adopted versions of the SBR.

Apart from the SBR relatively few anti-gas measures would stay the course and prove useful against most

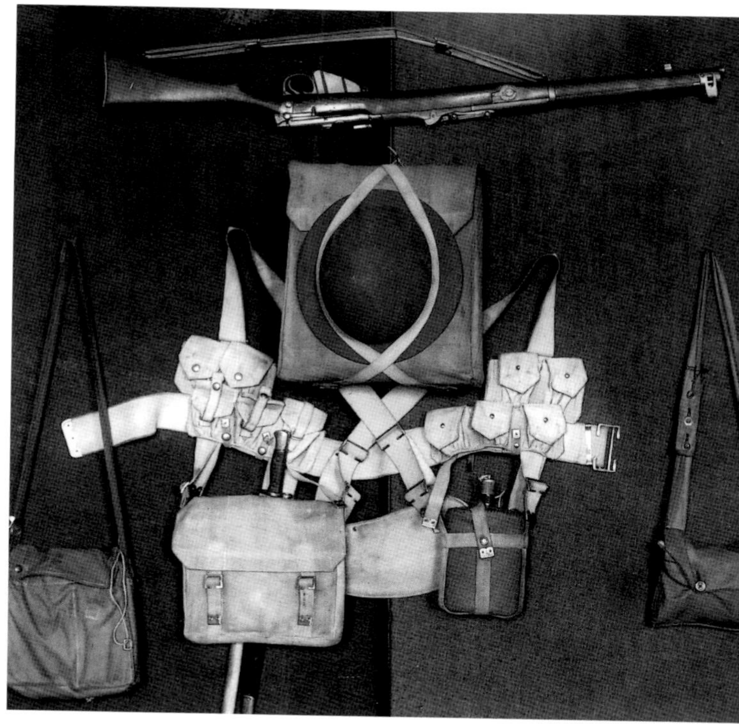

types of gas. A humble exception was the gas blanket which became a common feature across dugout doors. Sprays and decontamination chemicals would also remain in use. Fans proved a sometimes farcical failure. One such was the "Ayrton" fan, designed by Mrs Ayrton, intended to disperse and waft away gas concentrations. These huge and ridiculous fly swats did apparently reach France, but it is difficult to imagine that they saw much actual use.

Generally speaking gas warfare did not achieve the great things which had been expected of it. There were localized advances and sporadic casualties attributable to its influence, but it was not the wonder weapon which some of the scientists and generals on both sides had predicted. One of its most useful applications turned out to be the denying of ground to the enemy, at least temporarily, as was the case in the final German retreats of the war. In the British case the lack of offensive success was due at least in part to the evolution of enemy defensive measures. It was also true that British offensive activity was hampered by a lack of coherence in research, development, and management of the offensive chemical warfare program. Even so the balance between defense and offense would fluctuate as the specialists engaged a new gas, a new tactic or a new mask. Whether exaggerated or not, gas had introduced a new and strange facet to war which was not going to disappear.

CHAPTER FIVE
TANKS AND CAVALRY

The British High Command has been heavily, and in many cases justly, criticized for its inflated estimate of the value of cavalry. Even had the Battle of the Somme in 1916 or the Battle of Cambrai in 1917 led to a significant tear appearing in the German line it is doubtful whether cavalry in strength could have successfully crossed the resultant mess of shell holes and wire, and exploited victory in a meaningful way. While they were still on horseback a handful of snipers or a single machine gun could halt a whole cavalry brigade. Successively the Civil War, the Franco-Prussian War, the Boer War and the Russo-Japanese War had shown the diminishing worth of the man-horse unit as a weapon, and the increasingly anachronistic nature of the *arme blanche*, the sword and lance. While it remained possible to panic poorly armed and led natives in colonial wars with the thunder of hooves, it was a different story against steady riflemen.

Sir Douglas Haig seems to have been slower than most to appreciate the vulnerability of horseflesh to modern bullets, and, as Inspector General of Cavalry in India he had had a vested interest in the blinkered approach. His thinking at the turn of the century was that to increase the artillery and infantry at the expense of the cavalry was a "fatal error."

During maneuvers in 1903 he would complain to General Sir Edward Hutton that, "the umpires attach too much value to the killing power of the rifle, and constantly give faulty decisions, and force the cavalry to adopt a passive role." In Haig's scheme of things it was the infantry who would be assisted forward by the artillery, but the cavalry, who, sooner rather than later, would "exploit," and turn advantage into decisive victory. Even during the preparation for the Battle of the Somme Haig would tell his generals that the forthcoming "success" ought to be exploited "on the lines of 1806" when Napoleon's Marshal Murat engaged in a lengthy cavalry pursuit after the Battle of Jena. As his obituary in the *Cavalry Journal* would put it:

"We in the cavalry looked on Sir Douglas as our own... we know that until the day of his death he maintained that the cavalry maintained their value, and refused to acquiesce with any theorists who declared that there was no place for cavalry on the modern battlefield."

Such observations were deeply and tragically misinformed, but what should be remembered is that cavalry did still have a place in 1914, though that function was very different to its historic role. To put it crudely the cavalry horse was no longer a "charger," but a means of transport, and loath though many fox-hunting officers may have been to recognize the fact, the best use of horsemen was in reconnaissance, or as mounted infantry. The cavalry horse of 1914 was performing well if it could carry an observer to a point of vantage ahead of the army and bring him back unscathed, or carry a body of riflemen and machine guns to a place where they could lay an ambush, or deny observation to the enemy. These practicalities indeed were tolerably well appreciated both by the current divisional structures, and by cavalry training. A squadron of cavalry was allotted to each division, and brigades of cavalry were commonly employed as a screening force. The key was that the cavalry had to accept that it was not a battle-winning arm, but that by patrols it could improve the position and intelligence of the army as a whole.

Just such a patrol action was reported by Lieutenant Goodhart of the 20th Hussars, near Binche, in August 1914:

"On the way I saw three or four hostile patrols (each 25-30 strong) on each of my flanks, but none were within a mile of my route. On arrival at Seneffe, inhabitants reported that some German cavalry were at the level crossing and station, which was at the bottom of a fairly steep hill which formed the main street. At the top of this hill I halted, and saw below me at the station, which was 800 yards away, 50 cavalrymen dismounted. A hostile patrol was just starting

out up the hill, so I dismounted and gave strict orders that no one was to fire until I gave the order. The excitement was, however, too much for the trained 'scout' of my troop. He blazed off at the two points at about 300 yards, and missed. I then withdrew my patrol. When I got clear of the enclosed country I halted. A civilian on a bicycle coming from the direction of the enemy overtook me… Very soon afterwards I saw… a strong troop, about 25 men, mounted on dirty-colored horses… my signaller tried to semaphore them with two service caps, but I was not surprised when they took no notice. I therefore left my patrol, and rode over myself to 'liaise' with them. I got quite close to them before I realised they were carrying lances and were Huns."

Goodhart managed to outrun his pursuers, which was fortunate because he had forgotten to load his revolver.

Above: **British cavalry armed with both lances and rifles pass through a ruined village.**

He survived to make a report to his brigadier, having acted as the "eyes and ears" of the army.

It is worth remembering that, at the outset, there was little choice other than the horse for such duties. The army had precious little mechanical transport, and armored cars, which were still very much at the experimental stage, were certainly not yet capable of much by way of off-road performance. Occasionally the cavalry would grasp modernistic opportunities of their own volition. The 12th Lancers for example made good use of a captured German motor car, until it was taken away from them as "not issue." Other units made remarkable efforts to become motorized with, or

Right: **One of the old school of cavalry officer: Staff Captain H. Maddick 5th (Royal Irish) Lancers, served in France and at Gallipoli, died 1915.**

Far right: **Horses and bicycles: C Squadron, Duke of Lancaster's Own Yeomanry, cross the Somme at Brie Bridge.**

without, War Office help. The County and City of London Yeomanry ordered two Wolseley armored cars from their own funds, and the Westmoreland and Cumberland Yeomanry similarly purchased two vehicles on Isotta-Fraschini chassis. The 2d King Edward's Horse bought their own Talbot armored car.

Where the cavalry did come seriously unstuck was when they were used as a "shock" weapon. Even in the Boer War most of the cavalry had actually fought on foot, using the horse as transport. By 1914 attempts to charge the enemy on horseback were fraught indeed. The first such contact by 4th Dragoon Guards was a minor disaster, which the official history glossed over as being, "checked by fire." Private Ben Clouting's eyewitness view was more traumatic:

"It was a proper melée, with shell, machine gun and rifle fire forming a terrific barrage of noise. Each troop was closely packed together and dense volumes of dust were kicked up choking us and making it impossible to see the man in front. We were galloping into carnage... and there was utter confusion from the start. All around me, horses and men were brought hurtling to the ground amidst fountains of earth, or plummeting forwards as a machine gunner caught them with a burst of fire."

The 2d Cavalry Brigade was lucky to escape with 234 killed, wounded, and missing.

The 20th Hussars in the Great War records a tragicomic incident in which, while most of the regiment fought dismounted, their French interpreter attempted to dispatch a German with his sword from horseback. Three times the horse shied away until finally, in disgust, the Frenchman drew his revolver and shot the German. There were a few instances of British cavalry coming to blows with Uhlans and dragoons on horseback, but these were few and far between, and may have been as much a result of the protagonists' wish to fight that way, as its practicality.

As the war went on, the cavalry on the Western Front would see progressively less use as mounted men. Dismounted battalions were formed in 1915, and excepting minor differences in equipment, there was little to distinguish them in this role from the ordinary infantry. In 1916 the Household Cavalry provided the men for a dismounted Household Battalion; later they would also furnish the personnel for a siege battery. This was taken further still in early 1917 when 14 regiments of "Reserve" cavalry and "Reserve Squadrons" of the Yeomanry were amalgamated into just six regiments. Later the same year five yeomanry regiments serving as corps cavalry on the Western Front were dismounted and absorbed into infantry battalions. Thus it was for example that the 1st Ayrshire Yeomanry, and 1st Lanark Yeomanry became part of the 12th (Ayr and Lanark Yeomanry) Battalion, Royal Scots Fusiliers, while the 1st Duke of Lancaster's Own Yeomanry provided men for the 12th (Duke of Lancaster's Own) Battalion, The Manchester Regiment. While cavalry continued to serve, often very usefully, on other fronts, the correct but belated conclusion had finally been drawn that cavalry had no place among the trenches, gas, and machine guns of the Western Front. What really rubbed this home was not just the general shortage of men which made cavalry units look like useful pools of manpower, but the fact that the cavalry now had serious rivals for their old functions: Aircraft, balloons and motor vehicles for reconnaissance, and the tank for shock action.

Tank Development

Like so many inventions the creation of the tank was the work of many minds, and a novel combination of several pre-existing elements. Siege engines which used men or beasts as motive power had come into being in the classical era, as had armor; guns had existed since at least the early 14th century; the internal combustion engine was developing at the end of the 19th century; and tracks also existed at that time. Only in the Great War were all these elements brought together, and it would take a committee of enquiry to answer the question as to who had "invented" the tank. Indeed it remains easier to give an outline of that development, than to apportion the credit.

As early as 1907 a tracked vehicle manufactured by Hornsby and Sons of Lincoln, England, was tested by the British Army. It was as yet an inefficient beast and met with considerable opposition, informed and otherwise. The result was that the intended gun tractor was not adopted, and that Hornsby sold out its patents to the Holt company of California.

Though officially the thing was dead some soldiers carried on dreaming. Captain T.G. Tulloch imagined that a similar vehicle could be fitted with a metal "tank" to carry soldiers; the Austrian G. Burstyn, and the Australian L. de Mole put forward ideas for vehicles which were effectively "tanks" but were turned down. "Armored cars," that is armored vehicles with wheels, fared a little better, and a handful were actually built before the war. One of the first such was the Austrian Daimler armored car of 1904. In 1914 there would be further rapid development of this avenue, resulting in the reasonably practical and reliable Rolls-Royce armored car. This was backed up by Seabrook and Lanchester model vehicles, all of which served in Royal Naval Air Service armored car squadrons on the Western Front in early 1915. Wheels, however, would cope no better than horses' hooves with mud and

trench lines, and the armored car would see its most dramatic successes of the war in more fluid theaters of operation like the Middle East. On the Western Front, like the motorcycle and the truck, the armored car was essentially a rear area conveyance rather than the spearhead of attack.

It was at this point that two giants of the tank story would enter the frame: Colonel Ernest Swinton, Official War Correspondent, and Winston Churchill, then First Lord of the Admiralty (Britain's navy minister). Swinton felt, but was having difficulty convincing those that mattered, that the Holt tractors which were now about to be employed by the artillery would be better used in no-man's-land tearing up wire. Churchill already had under examination steamrollers and "Pedrails," track laying vehicles, which at the outset were animal rather than machine powered. The Army finally tested a Holt tractor over trenches at Shoeburyness in February 1915, but were disappointed its performance. The Admiralty now gained a clear lead, forming the "Landships Committee" under Eustace Tennyson d'Eyncourt, Director of Naval Construction, and including on it Colonel R.E.B. Crompton, an enthusiast of armored and powered "trench straddling machines." They continued to research and developed both the Pedrail, and a monster machine with huge wheels. Several experimental models would be produced, including the Tritton Trench Crosser, converted from a Foster-Daimler tractor, and a long thin and now mechanically powered Pedrail Landship. Another contender which entered the lists was the Killen-Strait tractor, effectively a large tricycle-like vehicle, but with three track laying units rather than wheels.

Unfortunately none of these put up a totally convincing show during trials, and not only was the Landships Committee now purged and reformed, but new areas of investigation were opened up. The main thrust was now at Lincoln where William Foster and Company, under the direction of William Tritton, was attempting to use its expertise in agricultural engineering to produce a workable war machine. Sent to work with them was a promising engineer, Lieutenant Walter Wilson of the Royal Naval Volunteer Reserve. Between them Tritton and Wilson would come up with the "Number 1, Lincoln Machine," an ungainly but, by the standards of the time, practical looking riveted steel box riding on a pair of American Bullock tracks. Part of their plan, not brought to fruition at this stage, was the mounting of a surprisingly modern turret with a 2-pounder pom-pom gun.

Number 1 was let down by weak tracks, but by November 30th, 1915 a partial rebuild and modifications to the tracks and track rollers produced that landmark in tank development – *Little Willie*. The machine certainly worked, and might have been refined, armed, and produced had not the parameters of the exercise been shifted at this stage. The War Office now demanded an 8-foot trench crossing ability, and a 4 foot 6 inch parapet climb, something which *Little Willie* certainly could not accommodate. It was a difficult hurdle but the designers came up with a simple solution which was to give most First World War tanks their distinctive shape. What they did was make the track frames larger than the hull of the vehicle and, to provide a significant climb ability, they then altered the shape of the track frames to a huge regular rhomboid.

Mother as the result became known, was a remarkable hybrid, with a number of features, which, for a land vehicle, seemed to have little historical parallel. Apart from the outsize and strangely shaped track frames the turret armament was abandoned in favor of sponsons, overhanging boxes of naval inspiration which gave the guns arcs of fire to the front and side. Quaintly production vehicles were to be either "Male" or "Female," the former having two 6-pounder guns and four machine guns, the latter having five or more machine guns. The vehicle had several methods of turning: Either by braking on one track; or by turning a wheel which turned some spindly looking trailing wheels; or alternatively by putting one of the track drives into neutral. The monster weighed about 28 tons and had a crew of eight. With a width over 13 feet, and a length just just over 26 feet, not counting the steering tail, maneuver was about as easy as steering a large domestic living room. Top speed on fair ground was a steady walking pace.

It is interesting to note that, although War Minister Kitchener was inclined to dismiss the new prototype as a "pretty mechanical toy," the High Command and Ministry of Munitions took the project seriously and were inclined to grant it significant resources. An initial order was placed for 100 of the Mark I, quickly upped by a further 50. After toying with other titles it was decided that the new tank unit should be known as the Heavy Section, Machine Gun Corps and that Colonel Swinton should command. Both production of the new vehicles and training men took time and, though Haig was keen enough to be pressing for tanks in the field as early as April 1916, it would be some time before any significant number was ready for action. Here indeed was a critical dilemma. One option was to bring the new armored corps onto the Western Front as rapidly as possible, even though this might mean that insignificant numbers were deployed and surprise was lost. The other possibility was to wait until tanks were available in large numbers, but to wait meant that the secret might be compromised, and it left the troops fighting and dying while a machine which could materially improve their chances was left out of the battle.

Tanks in Battle

Ultimately tanks were not ready to be committed to the Somme on July 1st but Haig decided that as many as possible should be thrown in to a renewed offensive on September 15th, 1916.

Getting the tanks to the right place at the right time would be a logistic nightmare. Since no road transporters yet existed the tanks had to travel by rail to the battle front and, since they were too wide for the standard rail cars, the sponsons had to be unbolted and carried separately. A Company would arrive too late to take any part in the action, and of the 60 tanks available 49 were in working order on the eve of the attack. Just 36, of C and D Companies, would make it to their start lines, between Thiepval and Combles. Their would task

Far left: **Personal equipment of an officer of the 20th Hussars: Breeches and non-regulation brown cardigan, spurs, crop, manuals, and the cavalry sketching board which, in theory at least, could be used one handed while attached to an arm.**

Left: **Cavalry cross a trench bridge near Neuve Église as Australian signallers pass underneath, May 1917.**

be to act in small groups against enemy stongpoints in the line and smooth the passage of the infantry forward. One of these vehicles deployed near Delville Wood was *Dracula*, commanded by Lieutenant A.E. Arnold of D Company.

"It was only three miles or so from Green Dump to my particular spot on our front line and I have always been puzzled why it took nine hours… to traverse that distance, for we seemed to be travelling the whole time and certainly made no deliberate stops. But it was a case of bottom gear all the time, and on good ground bottom gear only produced a speed of about 1,000 yards per hour. But it was necessary at times for one of the crew to get out and scout for a way round a particularly bad patch.

As dawn began to show the ground up we were still far behind the front line and it seemed certain we could not get there by zero hour for the going was now simply one succession of shell craters. But the ground was dry and it was thrilling the way the tank would go down into a crater, stick her tracks into the opposite wall and then steadily climb out. The rate of progress was now desperately slow and I suppose the last thousand yards took two hours to cover. We were getting nearer to the front line and although the infantry were out of their trenches before we arrived we now made better time…

We were into the German counter-barrage. Not all the tanks that left Green Dump ever reached the front line; some developed mechanical trouble and others became ditched. As we crossed no-man's land the other tank of my pair was just in front and a little way to the right and getting along well now that the ground was better. Suddenly she was stopped and emitting clouds of smoke. I saw the crew – or some of them – tumble out of the back door. This was not encouraging.

It was now half light; we were getting along better and among the infantry who were in turn advancing and sheltering in shell holes as our creeping barrage gradually lifted. The German shelling was severe and one felt comparatively safe inside the tank. The German front-line trenches had been shelled practically out of existence and I think the infantry met little opposition there. And Dracula reached the support line first. A row of German heads appeared above the parapet and looked – no doubt in some amazement – at what was approaching out of the murk of the bombardment. At point blank range I drew a bead with my Hotchkiss [machine gun] and pressed the trigger. It did not fire! Again I fired, and again the same result. But those inquisitive Germans gained only a momentary respite for the tank was on top of the trench and there we paused while the Vickers guns raked the enemy to port and starboard."

If jolting about inside one of these airless and overheated contraptions left the crew exhausted and disorientated, it was as nothing to the feelings of the German infantry on the receiving end of these "iron devils" for the first time. Relentlessly the tanks came on. In their path one of the enemy units was 17th Bavarian Infantry Regiment; small arms fire proved useless against the tanks and many men simply fled. Even in this initial attack, however, the tanks had not proved invulnerable: Ten machines were hit in one way or another, and seven were damaged. According to Gustav Ebelshauser's account at least one was halted by a close assault with hand grenades. As the tank was born, so was the specialism of anti-tank work, and thus began the search for effective anti-tank weapons. Yet the first tank attack was at least a limited success, a good bag of prisoners was taken, and it is very doubtful that the infantry would have got as far as they did that day without tank support. In the most celebrated instance one machine, D17 *Dinnaken*, got as far as the village of Flers, as was reported in Lieutenant Frank Mitchell's *Tank Warfare*:

"One tank led the way into Flers, the Germans flying before it in terror. Most of them bolted into cellars, and the New Zealanders, who were following simply had the job of rounding them up... A low flying aeroplane, acting as observer that day, sent back the message which was printed in every newspaper in the kingdom – 'A tank is walking up the High Street at Flers, with the British Army cheering behind.'"

Impressive though this was, it was a modest tactical success, rather than a general breakthrough, or a resounding victory. Sometimes the explanation given is that the High Command did not realize the full

Far left: **Rolls Royce light armored car, Abbeville, May 1916. (IWM Q 538)**

Left: **A Mark I tank on its way to attack at Thiepval, September 25th, 1916. The trellis-like structure on top supports anti-grenade netting.**

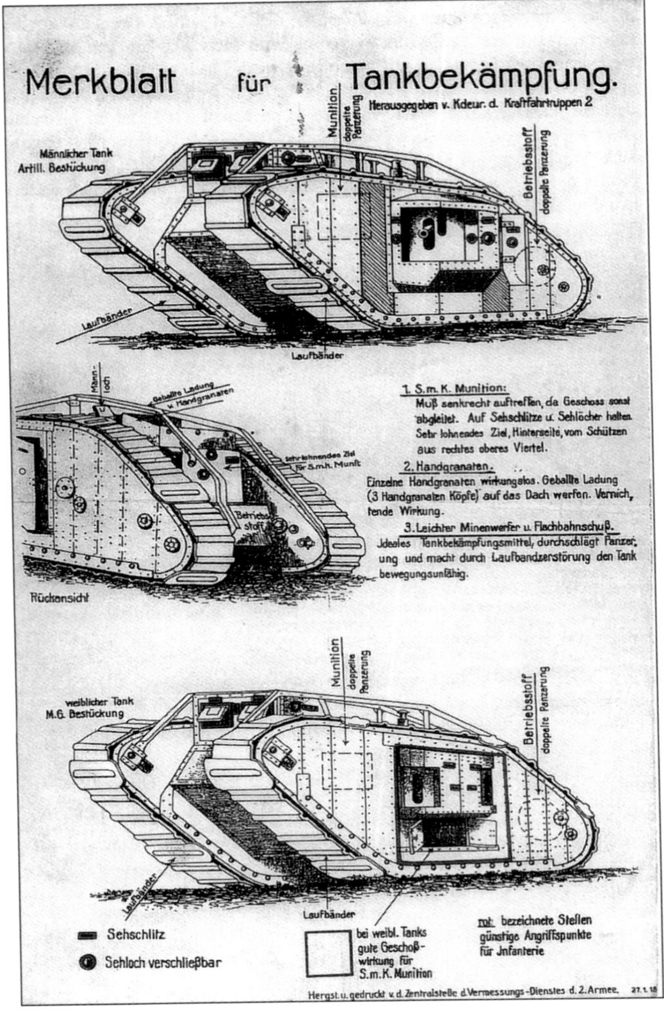

potential of their new weapon, or that its advantages were prematurely squandered on a relatively small scale enterprise. While there may be elements of truth in both these points of view it has to be said that the Mark I was not a breakthrough weapon, nor was it capable of sweeping independent action. Apart from questions of mechanical reliability and atrocious ground conditions, the tank had first been conceived as a sort of siege engine, capable of breaking into the enemy position, but not breaking out into the green fields beyond. It carried 50 gallons of fuel, enough for a radius of action of about 23 miles, perfectly adequate for a limited attack, but not enough to clear the enemy rear in a deep position. Refuelling was problematic until supply tanks could move forward with an attack; until then cans of fuel would be dependent on man and horse power to shift them across the devastated battlefield. With limited range and tenuous supply a tank commander's obvious course of action was to aid the infantry forward, dominate the enemy trench system and knock out machine-gun posts until low on fuel and ammunition, and then fall back. Neither was the tank invulnerable and shell fire could easily smash the relatively thin metal boxes.

Another obvious factor which precluded a deep and decisive tank breakthrough was the lack of communication which bedevilled almost every action taken on the First World War battlefield. Tanks were soon equipped with a basket of pigeons with which to send reports, but this was at best uncertain one-way traffic. Sometimes, for want of space, the pigeon basket was balanced on top of the engine and overlooked in the heat of battle. The birds emerged cooked, or as Frank Mitchell put it, "overheated and semi-asphyxiated." The next idea was semaphore arms on the roof of the tank with which the crew could communicate locally in a rudimentary fashion. This may have been of limited use to supporting infantry, but the signal had to be seen to be understood and was no way to communicate with command. Simple signs were also laid out on top of the tanks to give reconnaissance aircraft an idea of what was going on. By the end of the war radios were being fitted to some tanks, but these were specialist vehicles, and the fighting tanks would have to follow the cue of their colleagues, or send crewmen out to liaise with them.

Almost always, and until the war was nearly over, the best tank attacks were those which followed simple schemes to relatively limited objectives. Once the tasks had been set it was next to impossible to accommodate a significant change of plan, and the eight men inside the tank were pretty much on their own. Many of the tank commanders' accounts of battle read as attempts to justify their course of action, whether it was to carry on to almost certain destruction, or to turn around and head back. The decisions were theirs, and the information on which to decide came through a tiny hatch or mirror, the fragmentary views which Mitchell would describe as "fleeting glimpses" over gun sights, senses dulled by the "thumping of engines" and the roar of the guns.

While Haig remained convinced that he would need his beloved cavalry for the real work of exploitation the British High Command was not blind to the merits of tanks, and made more constructive use of them than any other protagonist. Within four days of the first tank attack 1,000 new machines would be requested with an immediate interim order for 100 in which to train the crews. In six months the tank had come from a bizarre idea, to a major officially backed scheme for the breaking of trench deadlock.

While new vehicles were under construction the original three companies of Mark I tanks continued to lend support to a series of lesser actions in France. Yet as their numbers depleted, and weather worsened, the Germans seem gradually to have gained confidence in dealing with the new menace. Artillery began to be deployed firing directly against tanks, and close action by the infantry became more common. Spirited resistance was encountered by Lieutenant H.W. Hitchcock of A Company, during an attack on November 13th, 1916,

"At five minutes before zero hour the engine was started and at zero hour Car [i.e Tank] No 544 advanced and was directed by Lt Hitchcock on its course till about 7 a.m. when it reached the German front line and was temporarily unable to proceed as the tracks would not grip owing to the condition of the ground... Up till now none of our own infantry had been seen and the car was surrounded by the enemy. About this time Lt Hitchcock was wounded in the head and gave orders to abandon the car, and then handed over command to Cpl Taffs. Three men and Lt Hitchcock got out of the car; Lt Hitchcock was seen to fall at once, but no more was seen of two of the other three men who had evacuated the tank. The third man was pulled back into the tank after he had been wounded in the forearm and, as the enemy were shooting through the open door it was immediately closed. Fire was at once opened on the enemy who retired to cover and opened on the tank with machine guns and rifles. Cpl Taffs decided not to abandon the tank but decided, with the help of the driver, L/Cpl Bevan, who had been previously wounded about the face by splinters from his prism, to carry on and get the tank forward to its objective. They managed to extricate the tank by using reverse and then drove forward as far as the German second line where the tank crashed into a dugout and was hopelessly engulfed and lying at an angle of about 45 degrees, thereby causing the two guns on the lower side to be useless and the two guns on the upper side only capable of firing at a high angle. The tank was now attacked by Germans with machine guns and also bombed from the sides, front and underneath... "

While the fighting at the front got harder great things were underway at home. Towards the end of 1916 the Mark II, and Mark III training tanks began to appear and were delivered to the new training area at Bovington, Dorset. Production of the thousand fighting tanks was also begun. The new model, known as the Mark IV, would ultimately become the commonest tank of the war, and though it had a good deal of similarity with its Mark I predecessor it included several refinements. Most obvious was its lack of external steering wheels, which had been found to have been more of a liability than an asset. The 6-pounder guns on the male tanks were also made shorter and more handy. A change was made to the shape of the sponsons, which were smaller, and now mounted Lewis

Far left: **German Second Army instruction leaflet dated January 1918, showing the areas of British tanks vulnerable to attack by armor-piercing "K" ammunition, grenades, and trench mortars.**

Left: **The tank was not invulnerable. Germans pose with the wreck of "Ghurka" a British tank smashed by shell fire.**

Left: **New Zealand artillery officers with a German 13mm, bolt action, single shot, Mauser T-Gewehr or anti-tank rifle. This was the first weapon of its kind. (IWM Q 11264)**

Below right: **A picture the censor would not let the British population see. The fate of many tank crew.**

guns rather than Vickers or Hotchkiss models. The smaller sponsons did not require unbolting for rail transit, and thus improved the tank's long range portability, but the Lewis guns' large barrel casing, as well as the tank engine's propensity to draw back fumes, were no advantage. One thing which was a significant advance was the addition of an "unditching" beam, a simple baulk of timber and chains carried on the top of the vehicle. When the tank got stuck this would be attached to the tracks, which would drag the beam underneath, and often provide sufficient traction for the tank to drag itself out of trouble.

Unfortunately the new tanks were not available for the Battle at Arras in April 1917, and so tank participation was limited to 60 machines. Even so these were now reorganized into battalions rather than companies, in preparation for the forthcoming expansion. In the event the role of the tanks at Arras would be disappointing, partly because of the Germans' increasing anti-tank proficiency. Armor-piercing bullets were employed, capable of breaching the tanks' slender armor, and deep anti-tank pits were dug, which, unlike a narrow trench, formed effective tank traps. A German report, prepared by 27th Division in April 1917, outlined both the tanks' perceived capabilities, and the measures which defenders could take against them.

"The machine guns at the fore end of the tanks open fire when within 500 to 100 yards of our lines. The guns of the male tanks can only fire to the front and side. Their arc of fire is considerable. On reaching or passing our trenches the majority of the tanks turn to the right or left, to assist the infantry in the mopping up of the trenches. Odd tanks go ahead to enable the infantry to breach our lines.

Ordinary wire entanglements are easily overcome by the tanks. Where there are high, dense and broad entanglements, such as those in front of the Hindenburg line, the wire is apt to get entangled in the tracks of the tanks. On April 11th one tank was hopelessly stuck in our wire entanglement. Deep trenches, even eight feet wide, seem to be a serious obstacle to tanks.

At long ranges by day, tanks will be engaged by all batteries that can deliver fire with observation and that are not occupied with other more important tasks. All kinds of batteries put tanks out of action on April 11th. Battery commanders must be permitted to act on their own initiative to the fullest possible extent… April 11th proved that rifle and machine-gun fire with armor-piercing ammunition can put the tanks out of action. Fire directed at the sides of the tanks is more effective than fire at the fore end. The greatest danger for the tanks is the ready inflammability of the fuel and oil tanks. Machine-gun fire is capable of igniting them. The garrison of the trench will take cover behind the traverses and will direct their fire at the hostile infantry following the tanks; firing on tanks with ordinary small arms ammunition is useless.

Anti-tank guns are indispensable; they are particularly useful for combating tanks which have penetrated our lines and are within our front lines; however anti-tank guns are a source of danger to our own infantry…"

The Mark IV tanks started to arrive in France in May 1917, and in June the huge expansion of the tank force was marked by the change of title to the Tank Corps. Though the Mark IV was no faster than its predecessors, the appearance of supply tanks, and the fact that the new vehicle had a larger fuel tank did suggest that the tactical role of the tank could be expanded. There was also a marginal increase in armor thickness; in 1917 would begin the race of armor against armor penetration which would characterize the development of armored vehicles through the remainder of the century. Hopes were high when tanks were committed to their next big action at Third Ypres in July 1917. Yet the enemy were not standing still, apart from having better anti-tank preparations their defensive positions were ever deeper, and the British were now facing concrete pill boxes, and a mass of shell holes and gun pits rather than simple linear defenses. In short, technical and tactical advances were running at such a pace that even a lapse of two or three months could change what had appeared to be the certainties of the stalemate. Trench deadlock may not yet have been broken, but the war was only standing still in a geographic sense; the specialists on both sides were moving fast.

To make matters worse the Passchendaele area was one of the least suitable for armor on the Western Front. Naturally waterlogged, the terrain was littered with the remnants of shattered woods and streams. Tree stumps and ditches could halt a vehicle while the enemy batteries homed in on the floundering beast. Some of the fiercest fighting was in the region of Frascati Farm, as was related by Lieutenant Maelor-Jones of G Battalion:

"At dawn on 31 July I proceeded with the other tanks of No. 3 section, crossing over the enemy front line in a north-easterly direction for Kitchener Wood to the right of the Oblong and Juliet Farms. At the latter I waited for our barrage to lift, meanwhile filling up our radiator with water. I then steered for Alberta, leaving it on my left. Here I met

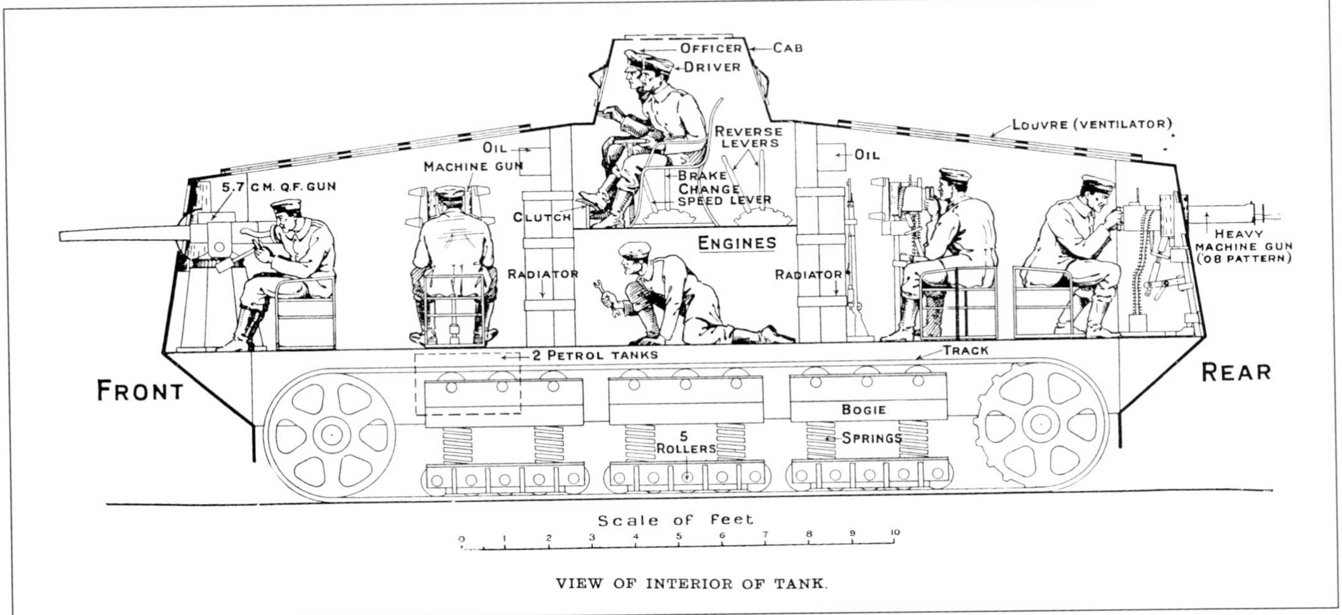

Above: **The German A7V tank. Though mobile fortresses, the German tanks were too little, too late, with only a couple of dozen appearing in 1918.**

some Hampshires retiring. With difficulty I persuaded a man to speak. He told me they were held up by machine guns. I advanced towards a long block house about six feet high where I observed a machine-gun emplacement on which I drove the tank. On the other side I came on a machine gun and two men whom we shot (one of them with a revolver through an open hatch) and crumpled up the gun."

This success was short lived. The tank was soon ditched and like so many others it was abandoned, and the top camouflaged with mud in an effort to deceive enemy airmen.

Dreadful though the offensive was, it did enable the tankers to use several techniques which would become important in the future. One of these was the deployment of supply tanks, which, on occasion, were able to rendezvous with the fighting tanks and refuel them. Cable laying tanks were also experimented with, the idea being that cables could be dropped off allowing the signallers and engineers to lay landlines for communication with the assaulting forces. Perhaps most remarkably Gun Carrier vehicles were used during September 1917, which, while retaining tank chassis and tracks, were loaded with 6-inch howitzers or 60-pounder guns and ammunition. The 6-inch could actually be fired from the tracked mounting. This idea was interesting because it seemed to offer a way to transport the heavy artillery across the trenches and the devastation that they themselves had caused. It also laid the germ of an idea which in future years would lead to self propelled and assault guns.

Below and Below right: **Two views of German A7Vs on the Western Front, June 1918. Note side hatch and machine guns.**

The scene was now set for the tank's most celebrated action of the Great War, Cambrai. Here the Tank Corps would field 378 tanks in the main attack, with a further 54 held in reserve. In preparation a huge concealed tank park was established at Plateau Siding near Albert, the tanks and crews being moved in by railroad transporters. Many of the tanks were equipped with fascines, great bundles of wood, weighing about 1.5 tons, which they would drop off to fill in ditches. Special wireless (radio) tanks and wire cutter tanks would also be used, in a systematic attempt to clear the path for the infantry, and maintain communication with the attacking units. Brigadier General Hugh Elles led the attack in person, flying the brown, red, and green flag of the Tank Corps from his own machine *Hilda*. As usual the costs were high, and after the battle, in one salvage dump alone, 16 carcasses of smashed tanks were to be seen with gaping holes blown by shell fire. A Victoria Cross was won by a tank officer who left his vehicle and attacked the enemy with a Lewis gun, a gearsman won a DCM for a similar action armed with only his revolver.

Curiously, Cambrai, unlike the great slogging matches of the Somme and Third Ypres, had never been planned as a protracted offensive. Thus it was that, although a significant advance was made, penetrating the enemy front to a depth of over six miles, the success was short lived. German counter-attacks quickly regained most of the ground taken. It was almost as if the tank was now being used as part of the wearing down battle, and an instrument of attrition. For tank officer J.F.C. Fuller, and other gurus of the potential of armored mobile action, this was not only wasteful but an indignity. Nevertheless Cambrai was accepted as a token of the tank's continuing worth, and was sufficient of a success to be represented at home as a significant victory. The tank arm would undergo further expansion, a Mark V tank would go into production, and plans would be laid for smaller, faster vehicles, the Whippets designed by William Tritton of Fosters.

Tank training was now further improved and systematized, with *Instructions For The Training of the Tank Corps in France* being published in December, 1917. The syllabus for the corps was divided up under four headings: Physical, moral, technical, and tactical. It is interesting that, even in what was plainly regarded as such a technical and specialized unit as the Tank Corps, moral training should receive such a high profile. Yet it was clear that, in the Tank Corps just as much as in the infantry, discipline, and leadership were crucial. The eight men shut up inside their steel box needed as much will power to go forward, and even more *esprit de corps* to act as a team. Each man was a specialist, but gunners, gearsmen, driver and commander had to act

as a unit, a unit indeed of shock action, and in this they had much more in common with the old fashioned cavalry than they might have cared to admit. Mitchell believed that tank crews were braver in practice, not through any predisposition to valor, but because being shut away, like "ostriches in the sand," they had little conception of the danger outside. Not having a regimental history with which to inspire the men was, however, keenly felt, and much was done to construct one in as short a time as possible.

No less than five separate schools now existed catering for Tank Corps personnel. The Gunnery School dealt not only with the 6-pounder main armament, but with the Hotchkiss machine gun, which had been brought back in again, and the revolver; the Mechanical and Driving School dealt worth both vehicle control and maintenance; the Anti-Gas School, and the Compass School were self explanatory; and the Wireless Signalling School dealt with, but never entirely solved the problems of, communication. Although many of these subjects were common to other arms, Tanks Corps' instruction now dealt with the subjects as the tankers would expect to confront them. In the Gunnery School for example the 6-pounders were fired only up to 1,800 yards, and special practice weapons with an air rifle attached were provided for cheap and simple close range aiming. The tank crews' closest machine-gun range was the previously unheard of point blank distance of 30 yards. Vanishing and moving targets were a speciality. There was also a Battle Practice range, with up to four tanks on it at any given moment. Revolver training was given in trench situations as well as on targets. An individual's training program was intended to last 36 days, and apart from the subjects already noted would include less obvious matters such as sports, air photography, "Swedish games," "trench jumping," grenade throwing, and a night march. The schools were so organized that they even informed pupils how many spare sets of underclothing, and which notebooks to bring.

The new instructions on tank training coincided with the publication of *The Training and Employment of Divisions*, January 1918, which outlined both the use of tanks, and the methods of their co-operation with other arms. Tanks were here defined as secure against "bullets, shrapnel and most splinters" and of great offensive value, but their limitations were also clearly expounded. It was not, for example, realistic to expect them to make a "circuit" of more than ten miles, nor could a speed in excess of 120 yards a minute be expected. At night perhaps only 15 yards a minute was practicable, and swamps, thick woods and the like were totally out of the question. When tanks co-operated with artillery it was expected that the guns would do four things in order to fit in with a tank assault. These were: Counter-battery work, especially against anti-tank guns; covering by means of smoke; and timing barrages so as to fit into the tank program; lastly, and perhaps most importantly, artillery was to avoid cratering the route of the tanks. Tank attacks could take place with or without prior artillery preparation. In co-operation with infantry, close liaison was seen as critical, infantry being on hand to exploit opportunities created by the tanks, and tanks helping to place the foot soldiers "securely on their objective."

In terms of formations and frontages it was envisioned that one tank would be employed every 100–200 yards of the attack front. Not more than one company of 12–16 tanks would be allotted per objective, and usually sections of four were kept intact. Four or eight independent tanks would move ahead of the formation as an advanced guard, while the rest covered the infantry, who followed them, in elongated section single files. On coming upon strong points the tanks would engage, forcing the enemy inside and under cover, holding and distracting them until the infantry could come up to make the final capture. Attacks could be made without preparatory wire cutting in which case the tanks themselves would be tasked with wire breaking. At the actual moment of wire crossing the infantry were well advised to hang back to avoid snapping and dragging wires.

Simple signals between infantry and tanks were standardized. A green disc shown by the tank signified to the accompanying infantry that the wire was cut, a red disc that the wire remained. Both discs shown together indicated that the tank had reached its objective. When the infantry wanted tank assistance they were instructed to hold a helmet or rifle above their heads. Two pigeons were still carried by each tank for long range communication. "Wireless signal tanks" were referred to in *The Training and Employment of Divisions*, but they were still thought of in early 1918, as "in the experimental stage."

Before any of this could be brought to fruition as attacking strategy, the Tank Corps, like the rest of the army, would have to weather the storm of the German spring offensive of 1918. Immediately highlighted was the fact that tanks, as tanks, had as yet no defined defensive purpose for a retirement situation. Usually therefore crews were dismounted and took advantage of their large numbers of machine guns to act as a bolster to the infantry. Unless acting as a stationary pill box the tanks themselves were best kept out of the way of the

Left: **Australian Engineers manhandle a dummy tank near Catalet, September 1918. Camouflage and deception became a major facet of trench warfare.**

attackers, who could run faster than the tanks could crawl. Brigadier Elles did hatch an extraordinary "Savage Rabbit" scheme in which tanks lurked in ambush and sprang out to cut off attackers, but it was generally a waste of tanks.

An exception to the rule were the new Whippets, with crews of three or four, armed with machine guns, which could manage about eight miles an hour, and had a significant radius of action. Perhaps the most celebrated fight in which they took part, was that of X Company, 3d Battalion, under Captain T.R. Price at Cachy, on April 24th, 1918. As Price would later report:

"I was lying up in a small wood near Bois de Blagny when I received orders from 58th Division, to whom I was attached, giving the information that an aeroplane had just dropped a message to say that a force estimated at about two enemy battalions, were massing for an attack about 1,000 yards east of Cachy and ordering my company to disperse them before they could launch their attack. It was a cold morning, and our engines were already running to warm up, so we were off at once. I led the tanks until we reached a small hollow where I left them and rode forward [on horseback] to reconnoitre.

Here I was delighted to meet Capt Sheppard of the Northamptons (my own old regiment), then commanding what was known as Sheppard's Force. He informed me that the first phase of the enemy attack on our line had worn itself out, but that he was expecting a renewal at any moment, and indicated the region where he judged the leapfrogging troops must be assembling. I ascertained that the country was beautifully open and undulating, and ideal for tanks, and galloped back to the company.

I assembled section and tank commanders quickly, gave them the information where I considered the two battalions to be, ordered them to form line, facing south at 50 paces' interval between tanks, cross Cachy Switch (no obstacle battered to pieces) and charge at full speed southwards, dispersing any enemy met on the way. On reaching the sky line which I indicated, they were to turn back and charge through the enemy again on the way back. My deductions as to the position of the enemy proved correct. The charging tanks came upon them over a rise, at point blank range, apparently having a meal served as several bodies had laid aside arms.

The tanks went straight through them, causing great execution by fire, and by running over many who were unable to get away. They turned and came back through the remnants again, utterly dispersing them, and the second phase of the attack on our line that day never materialised. We lost one tank, the commander in his enthusiasm crossing the sky line which I had indicated as the limit of advance, and being knocked out by a battery placed somewhere in the vicinity of Hangard Wood. What the total casualties of the enemy were is unknown, but 400 dead at any rate were counted later."

The German spring offensive also marked two other remarkable departures in tactical terms for tank warfare. The first of these was that for the first time the Germans now deployed their own tank, the monstrous 30-ton A7V, which had a crew of 18, and mounted a 57mm gun plus six machine guns. Fortunately only 20 were built, but the Germans also had access to larger numbers of captured British tanks, and thus it was, for the first time, that the British Army had seriously to employ anti-tank tactics of its own. The other important development was that for the first time tanks now met tanks in action, and the first faltering steps to what we might understand as armored warfare had begun.

That anti-tank measures would have to be employed had been at least considered as early as 1917, and in February 1918 a provisional document appeared entitled *Instructions For Anti Tank Defence*. This supposed both that it was entirely possible that the enemy would employ tanks, and that they might be faster than the British vehicles currently in use. No plan was then laid for the employment of specially designed anti-tank guns, it being assumed that "existing artillery resources" were adequate to the task. Though much of the artillery was to be devoted to destroying the infantry which would accompany enemy armor, it was appreciated that German tanks would have a considerable "moral effect" and should be engaged at the earliest opportunity. Remarkably it was accepted that artillery fire against tanks might cause friendly casualties, but asserted that the troops themselves would accept this, because the infantry "must realise that such losses cannot be compared to the damage and casualties which will be effected by unengaged tanks."

By means of vigilance, in terms of air reconnaissance as well as local intelligence, the artillery would seek to engage enemy tanks from first discovery. The guns would ideally be deployed in depth so that the armor would have a deep field of danger to cross before creating any breakthrough. With any luck defensive barrage fire would disable or discourage enemy tanks but, if this were not the case, the tanks would be engaged over open sights. Some guns would be deployed forward, dug in and strongly protected for this purpose, although it was not envisioned that there would be sufficient available to do this everywhere. The use of "mobile sections" of light field guns was encouraged, and where possible these were to use pre-selected fields of fire, where they might shoot down village streets, or across likely avenues of approach. High explosive was recommended for engaging tanks, since shrapnel was ineffective, and smoke obscured the target. In these respects the new British plan mimicked established German practice.

Above: **German Stormtroops, including a flamethrower operator, take the rare opportunity of tank support from a captured British tank.**

Infantry were seen as less effective in the anti-tank scheme, and instructed to reserve most of their efforts for enemy infantry. In confined spaces, however, they might use charges or mortar bombs with some success against tanks. Machine guns could give some direct help, and it was recommended that when tanks got really close several machine guns should concentrate on one vehicle. This would maximize the effect of any armor piercing ammunition available, forcing the tank crew to shut down their hatches, and possibly stop the tank crew manning their weapons. One special infantry anti-tank weapon was developed before the end of the

war, a rifle launched anti-tank grenade No 44. Though this can have seen little if any use against German tanks, it was realistic given the slender armor then in use, and set the tone for other similar devices in the future. Anti-tank obstacles were perceived as a useful part of anti-tank work. A high parados of soft earth, fallen trees, tree stumps, and specially blown craters were all recommended. A variation on the theme was to dig wide anti-tank ditches across roads and routeways; these could be lightly bridged so as to allow friendly troops to cross, but insufficiently spanned to allow the passage of tanks.

Interestingly tank-versus-tank action was also considered in *Instructions For Anti Tank Defence*. Apart from aiding the infantry to take back the ground it was postulated that individual tanks might operate as "mobile batteries to destroy individual tanks." In such a case it was envisioned that the British tanks would gain familiarity with the ground, and co-operate with the artillery in mapping out their own routes forward, so as to avoid becoming casualties to friendly fire. Tank-to-tank action was expected to take place first, and be swiftly followed by the infantry-supported tank counter-attack.

The mobile battery theory was put to the test with the first tank-to-tank battle at Villers-Bretonneux, in April 1918, not far from the scene of Lieutenant Price's Whippet action. Both sides would claim success. Lieutenant Frank Mitchell described the historic event in the following dramatic terms:

"Suddenly, out of the ground ten yards away, an infantryman rose, waving his rifle furiously. We stopped. He ran forward and shouted through the flap, 'Look out! Jerry tanks about!'. Swiftly he disappeared into the trench again, and Captain Brown immediately got out and ran across the heavily shelled ground to warn the female tank. I informed the crew, and a great thrill ran through us all. Opening a loophole I looked out. There some three hundred yards away, a round squat looking monster was advancing; behind it came waves of infantry, and farther away to the left and right crawled two more of these armed tortoises…

We still kept on a zigzag course, threading the gaps between the lines of hastily dug trenches, and coming near the small protecting belt of wire we turned left, and the right gunner, peering through his narrow slit, made a sighting shot. The shell burst some distance beyond the leading enemy tank. No reply came. A second shot boomed out, landing just to the right, but again there was no reply. More shots followed.

Suddenly a hurricane of hail pattered against our steel wall, filling the interior with myriads of sparks and flying splinters. Something rattled against the steel helmet of the driver sitting next to me, and my face was stung with minute fragments of steel. The crew flung themselves flat on the floor. The driver ducked his head and drove straight on.

Above the roar of our engine sounded the staccato rat-tat-tat-tat of machine guns, and another furious jet of bullets sprayed our steel side, the splinters clanging against the engine cover. The Jerry tank had treated us to a broadside of armour piercing bullets!

Taking advantage of a dip in the ground, we got beyond range, and then turning, we manoeuvred to get the left gunner on to the moving target. Owing to our gas casualties the gunner was working single handed, and his right eye being swollen with gas, he aimed with the left. Moreover, as the ground was heavily scarred with shell holes, we kept going up and down like a ship in a heavy sea, which made accurate shooting difficult. His first shot fell some fifteen yards in front, the next went beyond, and then I saw the shells bursting all around the tank. He fired shot after shot in rapid succession every time it came into view.

Nearing the village of Cachy, I noticed to my astonishment that the two females were slowly limping away to the rear. Almost immediately on their arrival they had both been hit by shells which tore great holes in their sides, leaving them defenseless against machine gun bullets, and as their Lewis guns were useless against the heavy armour plate of the enemy they could do nothing but withdraw…

As we turned and twisted to dodge the enemy's shells I looked down to find we were going straight into a trench full of British soldiers, who were huddled together and were yelling at the tops of their voices to attract our attention. A quick signal to the gearsman seated in the rear of the tank and we turned swiftly, avoiding catastrophe by a second.

Then came our first casualty. Another raking broadside from the German tank, and the rear Lewis gunner was wounded in both legs by an armour piercing bullet which tore through our steel plate. We had time to put on no more than a temporary dressing, and he lay on the floor, bleeding and groaning, while the 6-pounder boomed over his head and the empty shell cases clattered all round him.

The roar of our engine, the nerve-racking rat-tat-tat of our machine guns blazing at the Boche infantry, and the thunderous boom of the 6-pounders, all bottled up in that narrow space, filled our ears with tumult, while the fumes of petrol [gasoline] and cordite half stifled us. We turned again and proceeded at a slower pace. The left gunner, registering carefully, began to hit the ground right in front of the Jerry tank. I took a risk and stopped the tank for a moment. The pause was justified; a well aimed shot hit the enemy's conning tower, bringing him to a standstill. Another roar and yet another white puff at the front of the tank denoted a second hit! Peering with swollen eyes through his narrow slit, the gunner shouted words of triumph that were drowned by the roar of the engine. Then once more he aimed with great deliberation and hit for the third time. Through a loophole I saw the tank heel over to one side; then a door opened and out ran the crew."

The German machines broke off the battle. Mitchell was awarded the Military Cross for his "coolness and initiative," in this far from one-sided fight. German accounts of the actions of the *Sturm Panzerkraftwagen Abteilungen*, or tank detachments employed on 24 April, stress the bigger picture, and state that from their point of view they could see not three but eight British tanks. Tank number two of the Steinhart group commanded by Leutnant Biltz appears to have been that knocked out by Mitchell's guns, three hits on it were noted, one of which penetrated the right front, and another which smashed the oil tank. Another tank of the detachment, Leutnant Stein's *Elfriede* was ditched, and fell into British hands relatively unscathed. Also caught was at least one of the tank crew.

It was this fortuitous capture which formed the basis of the British intelligence document *The German Tank "Elfriede"* which was circulated in June 1918. Apart from confirming the armament and technical details of the engines, it allowed British engineers to measure the thickness of the armor and seek out the monster's weak spots. They concluded that all the 14 loopholes were potentially vulnerable, to "splash" and direct fire, that the floor was weak, as was the area in front of and behind the cab, and that an artillery shell virtually

Above: **Motorization for breakthrough: General Horne inspecting 24th Motor Machine Gun Battalion at Dieval, June 1918. (IWM 10325)**

anywhere could prove fatal to the tank. It was noted that observation from within was poor, neither the driver nor gunner being able to see points on the ground within ten yards of the tank, and the machine guns could be trained no closer than five yards. The information was useful, and to an extent comforting, but since so few German tanks were encountered from then on tank-to-tank tactics would develop little further in 1918.

When the German spring offensive had finally spent itself, with huge losses, allied power began to revive. This would be expressed not only in terms of American troops joining the fray, but in more and new tanks. The new Mark V was blooded on July 4th at Hamel, fittingly enough with both Australian and American infantry in support. Although similar in general outline, the new tank offered several advantages over its predecessors. Its armor was marginally thicker at 14mm, the cab was raised for slightly better visibility, and there was provision for an extra machine gun at the rear. Most importantly improvements had been made to the drive and controls. The driver was now pretty much capable of operating the machine single handed, rather than having to co-ordinate the action of gearsmen and brakemen. The result was a much more manageable vehicle, which allowed the rest of the crew to concentrate on the fighting and command.

Hamel would demonstrate that tanks could achieve a shock value, without preliminary bombardment to warn the Germans that they were coming. Just 60 tanks succeeded in taking the set objectives, in more than one instance running over and crushing machine-gun teams in their path. Armor-piercing bullets were used against the Mark Vs, but usually failed to do significant damage. Good use was also made of tanks as carriers for supplies, and it was estimated that they moved up to the objective 50,000lb of wire, iron sheeting, bombs, water and ammunition. Perhaps briefly, on a local scale, the tank had achieved superiority over defense, but the technological battle of machinery and tactics was a pendulum. Sooner or later the new German anti-tank rifle, the T-Gewehr, would become a serious factor in the equation. Perhaps even more significantly the radius of action of the new tanks, though improved, was still only 45 miles, and communication problems were still far from solved.

The Battle of Amiens, begun on August 8th, 1918, would see one of the greatest tank concentrations of the war. About 450 were involved, with the Mark V fighting alongside light Whippets and gun-carrier tanks. There was often a whole tank battalion employed per infantry division. In three days' fighting a six-mile advance would be achieved, which did actually succeed in pushing tanks across the enemy lines of communication. It may not have been a complete breakthrough but it was an effective demonstration of what enough of the latest tanks, combined with infantry, could achieve. It was also a milestone on the way to the final destruction of the German Army on the Western Front. Moreover tanks and infantry rolled forward without a preliminary bombardment; guns co-operated simultaneously rather than sequentially. Perhaps just as importantly, when the attack began to slow, it was discontinued; there would be no repeated battering to wear down the attacker as much as the attacked. Yet tank warfare was never easy, and casualties were often heavy with opposition from increasingly sophisticated forms of anti-tank defense, as the report on the action of *Barrhead*, tank number 9003 of 2d Battalion demonstrates:

"Strong opposition was met from machine guns, anti-tank guns, artillery and bombing aeroplanes. The machine guns were soon silenced, Barrhead's six pounder guns opened fire on some splendid targets and her machine guns poured forth a leaden hail of bullets on the Germans who were seen running in all directions. Pushing ahead and getting nearer the objective, the artillery fire became very heavy; shells kept bursting around Barrhead so the driver steered a zig zag course to avoid them and meanwhile the gun men kept up a heavy fire. At this time one of the crew was wounded, and while the NCO was examining his wounds, the tank was hit by a shell. The concussion from this shell threw the crew all over the tank and filled it with suffocating fumes. I got four of the crew outside and placed them at the rear of the tank as they were all wounded. On re-entering the tank to ascertain what had happened to the other two members of my crew I found them both dead. The shell, which must have been a large high explosive, had hit the tank in front of the right-hand sponson and burst inside wrecking the cylinders of the engine.

After dressing the wounded men I sent three of them to the nearest dressing station and went in search of a stretcher for the other man whose wounds prevented him from walking. While I was bringing the stretcher the tank was hit again and burst into flames."

Sadly such experience was far from typical, since attrition applied to tanks just as much as to infantry, and, given the limited number of machines, the number of attacks, even successful attacks, which could be made in a given time period was strictly finite. According to one recent calculation the last three months of the war saw a third of all tanks committed disabled in one way or another, although the vast majority of these were eventually unditched, or rescued and repaired. Tank Corps' personnel suffered in proportion. In the same period about a third of the crewmen became casualties, an average of two men every time a tank was committed to battle.

Tanks were one of the instruments with which the Hindenburg Line was finally broken in September and early October 1918, and among the vehicles deployed was another new model, the Mark V*, which was lengthened by about six feet, and thus had a considerably enhanced trench crossing ability. It would also not be too much to say that tanks, and the new specialists that drove and fought in them, were one of the weapons which would ultimately help the allies to gain the upper hand on the Western Front in the closing stages of the war. However, the tank was not a war winning weapon by itself, tanks working with infantry and artillery achieved these successes, and there were major factors outside the tactical arena of the battlefield, like the allied naval blockade, and the entry into the war of the United States which would make it arguable to what degree the tank had helped to turn the tide.

Whether the tank was critical to the outcome on the Western Front, or merely complementary, it cannot be denied that the Great War had given birth to a battlefield specialism which would have a remarkable impact on many of the wars of the 20th century. Industrial and technical effort had been concentrated on producing a new shock weapon to help break the stalemate, and the trenches had helped to bring forth one of the instruments of their own undoing. Eventually most of the world's cavalry would be transmogrified into armor. The "Iron Devils" of 1916 would have far reaching consequences.

CHAPTER SIX

FROM 1918, TO BLITZKRIEG AND BEYOND

The established picture of the British High Command on the Western Front is one of remoteness, incompetence, and lack of receptivity to new concepts: the popular view of tactics is that they scarcely changed. The butcher's bill was undeniably catastrophic: total British and Empire deaths, (including the foray into Russia in 1919) were 908,371; most of these were British, and the vast majority on the Western Front. Almost 2,100,000 were officially reported as wounded. Yet the popular stereotypes are at best gross oversimplifications. As we have seen the ways of war changed almost as much between 1914 and 1918 as they were to do between 1939 and 1945, and whilst there were some very bad generals there were some good. It was equally true that there were very bad politicians, and good reasons why Britain did not put every effort into the war from the outset. As Kitchener put it to Sir John French in a letter of August 27th, 1914,

"We are all determined to support you to the utmost, and to see that as soon as possible, you shall be provided with an adequate force, which will increase as we go on . . . Do remember that we shall have to go through much more fighting before we are out of the war, and that by prematurely putting all our eggs in one basket we might incur far greater losses. Believe me, had I been consulted on military matters during the last three years, I would have done everything in my power to prevent the present state of affairs."

Above: **Frank Brangwyn poster for War Bonds. As the war progressed it was perceived less as a matter of righting moral wrongs, and more as something to "get finished."**

What should, perhaps, be asked is whether the record of the British High Command in World War I was objectively worse than that of other nations. In terms of sheer slaughter the answer must be negative: the shock of loss to the British people was huge, genuine, and in modern times, unprecedented, but no worse than in Germany, France, and Russia. The wrench to Austria was arguably greater still, and spelt her end as a significant power. Even Italy who joined the war comparatively late had her costly "Caporettos." Where Britain lost her hundreds of thousands Russia, France, and Germany lost their millions. Neither Sir John French nor Sir Douglas Haig were brilliant generals, but they were not as profligate as Joffre or Falkenhayn.

Perhaps no British or Empire Great War general was truly great: but there was a very considerable spread of ability. "Daddy" Plumer, Monash, Maxse, Birdwood, Horne and others all showed signs of talent, or at least

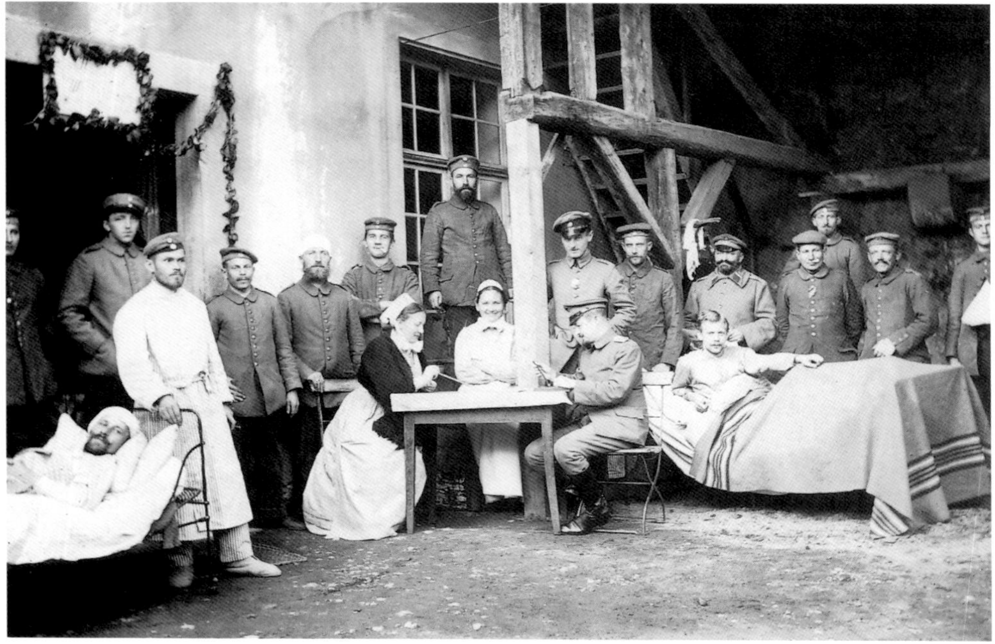

Right: **German wounded, November 1915. Despite huge losses the German Army would remain remarkably resilient.**

Below right: **An intelligence officer questions a wounded German prisoner near Contalmaison, July 1916. (IWM Q 805)**

common sense. Near the other end of the scale intelligence chief Charteris made ludicrously optimistic estimates of the imminent collapse of Germany and thus did much to encourage the continuance of battles which should have been abandoned. Kiggell, as Chief of Staff, was the very caricature of the "Chateau General," short on practical experience, and sadly out of touch with what was happening on the ground. Late 1917 and early 1918 saw the removal of Kiggell and Charteris, and even Sir William Robertson, Chief of the Imperial General Staff, but somehow Haig would cling on as commander of the armies in France.

Perhaps the reason for lacklustre performance at a senior level lies in the nature of the war itself. A few men did shine in the war, or were made to shine by the patriotic press, but almost without exception these were junior leaders, or specialists doing a specialist job. People like T.E. Lawrence in the desert, Von Richtofen, Albert Ball, and even Hermann Goering in the air, did what they did very well. Colonel Bruchmuller, or popularly "*Durch-Bruchmuller*" (Breakthrough Muller), the artillerist, similarly was a specialist. None were Army Commanders, and none were generalists. Those who would come to prominence in the future like Montgomery, or Rommel, were unknown in 1918. The "heroes" of World War I wore the V.C. or the Blue Max, led battalions, or patrols, flew aircraft, or wrote poetry, but never led armies. They were predominantly young, and many would never grow old.

Perhaps the war was moving too fast, and was too vast to be readily amenable to the decisive intervention of a single military mind. By the middle of the war GHQ was attempting to provide mathematical calculations of "wastage," and the "number of yards per gun," or shells per bombardment, and useful though it was to attempt a quantification all such figures were bound to be out of date almost before they were circulated. Once high explosive shells were available in quantity it was found that the Germans had dug in deeper so the shells and fuses needed were different. Once a way was found to break a single trench line it was discovered that the defence was digging its system in depth. Sir John French asked for trench mortars, for gas, for more shells, and for huge supplies of men. By the time they had all arrived he would no longer be in command.

In terms of the "bigger picture" the British Army and its generals on the Western Front were seldom masters of their own destiny. For one thing both French and later Haig were commanders only of their particular front: at home the Secretary of State for War, and the Chief of the Imperial General Staff, had the grasp of the war in general. These men were also fallible, and responsible to the cabinet. Though obviously an apologist for the actions of the High Command in France Brigadier John Charteris had more than a grain of truth in his observations when he wrote of 1915 that:

"If the cabinet had ever bestowed their complete confidence on Lord Kitchener, they had already withdrawn it, and would not accept his advice without endless discussion and criticism. Cabinet meetings degenerated into interminable debates on strategic plans. Schemes were advanced from all quarters, and each bore the common characteristic of inadequate knowledge of the limitations and requirements of modern warfare."

Among other things to flow from these top level discussions were Winston Churchill's championing of the Dardanelles, a scheme of Lord Fisher at the Admiralty for a landing in the Baltic, and a thrust through Greece and the Balkans at Turkey. Perhaps most extraordinary was Lloyd George's suggestion that all but a rear guard be withdrawn from the Western Front and the majority of the army be redeployed. Even Kitchener did not appear to have been totally committed to the Western theatre of operations, putting forward the idea that the British army in France might be regarded as in investing force, whilst real offensive pressure could be applied elsewhere.

Later Lloyd George and Haig would be at odds over Western Front strategy. Haig was instructed to follow the French lead, and to take pressure off Britain's ally in the wake of Verdun. The Somme was the result. In 1917 the Third Ypres was specifically approved by Lloyd George; but when it failed he immediately distanced himself from the result. When Haig died in 1929 Lloyd George could now safely blame him for every set back and blunder.

On the purely technological side there were also very severe obstacles to decisive campaigning. Away from the front trains and ships could carry armies long distances at reasonable speed: this was never the case at the "sharp end," even during the comparatively open periods of warfare in 1914 and 1918. On the battlefield everything moved at walking pace. Even if a Division could be planted in reserve three miles from the front line ready to reinforce an attack, it would probably take at least a couple of hours just to reach the point where the battle started. Perhaps half the day would have elapsed before it reached the point to which it had been directed, and for all this time it would have been within range of enemy artillery. Any change of plan would throw all calculation out of balance.

Communications were another apparently insuperable problem. In 1914 many messages were carried by runners, riders, and pigeons, waved with flags, or even flashed by heliograph. There were relatively few telephones, and even moving at a mile or two per hour armies were apt to run ahead of their phone lines. The year 1915 saw a massive increase in the number of telephones, but whenever an army attacked there would be a critical period in which reliance on runners and waving flags left the Generals in ignorance. Defeats would be senselessly reinforced, and potential breakthroughs missed.

With the trench line ossified telephone problems were largely mastered in 1916. New grids were laid out in standardised patterns, but shelling could, and frequently did, severe communications at the vital moment. At the same time signals traffic was diversifying as the specialists got busy. Messenger dogs, lamps, buzzers, rockets and klaxons helped to create what to the nineteenth century mind must have been a cacophony of sound and text. Now the headquarters suffered from information overload as messages from different parts of the battlefield arrived at widely differing speeds.

The next twelve months would therefore see innovations regarding signal silences, signals security counter measures, and advances in wireless. Though much improved the communications network of 1918 was still imperfect. Wirelesses were still huge delicate beasts requiring clumsy batteries, and whilst some vehicles were capable of carrying them there were as yet no man portable sets for the infantry. Senior staff officers are often criticised, sometimes correctly, for not looking at the conditions the men had to face first hand, but if a senior officer left his HQ, or his telephone lines, he became temporarily useless. Haig did not improve the situation by expressly forbidding staff officers from going to close to the front, because they were difficult to replace if killed. Contrary to the popular picture 49% of pre war trained staff officers would be killed or die of wounds, a disproportionate number of them by the end of 1915.

Contrary to general belief many important lessons were learned from the debacle of the Somme, and more from the mud of 1917. At the tactical level the art of attack came a long way: barrages were no longer simply timed batterings but complex fire plans which could be mapped, and calculated to lift, creep, and throw smoke or gas. The troops on the ground were now commonly fronted by tanks, and better placed to leapfrog from shell hole to shell hole because their vanguard was usually fronted by light machine guns and bombers. Rigid waves of attacking troops began to give way to more flexible formations.

Detailed training for the attack, and for tactical combat, became the norm rather than the exception. The men were trained both in specialisms, and the unique

features of their specific task in a specific battle, in a way which was ever more realistic. As one account of the attack preparation of a battalion from November 1916 put it,

"The attack itself was practised, whenever dry ground could be obtained, over dummy trenches. New ideas of 'battle order', bomb carriers, distinctive badges etc., were dished out to the men till everyone was heartily sick of the battle long before it began."

It has been argued, from a revisionist stance, and in reply to the "mud, blood, and endless slaughter" model, that the trench warfare was actually more of a "live and let live" system. This may be true at some times, and in some places, since there were temporary unofficial truces, and fire sometimes became "ritualised." There were sectors for example where there were "morning hates" directed with regularity, which men learned to avoid. These glimmerings of humanity were however the exceptions to the general rule. The trenches existed to protect, and protect they did, but the thrust of the efforts of the men in them were to find or employ new ways to either kill the enemy in his emplacements, or drive him from them. Generals dictated a policy of aggressive raids and patrolling; machine gun fire at night denied movement to the enemy: snipers; grenades; bomb throwers and trench mortars ensured that the trenches never became a haven of peace. What was more these instruments of destruction were not ossified, but were constantly being updated, developed very often by the front line soldiers themselves. This was truly a soldier's war, and a war of "specialists."

Remarkably "open warfare," the holy grail which both sides had grasped for over more than three years, would result in more death over a given period of time than grinding away at the enemy trench lines. The bloody 1916 battle of the Somme lasted about five

Above: **The empty battlefield. A well concealed German patrol have camouflaged their steel helmets with foliage, disappearing into the landscape, c.1918.**

Far Left: **The new technology reaches entertainment – soldiers at the "British Cinema." By 1916 even the public at home would see graphic footage with the general release of the official film *Battle of the Somme*.**

Left: **Trench raiders of the Liverpool Irish, 1916. Raiding helped to develop new tactics, gained control in no-man's-land and was also used as a way of preventing units from becoming stale or apathetic.**

months: including the wounded and missing 498,000 casualties were counted. Third Ypres, including the battles commonly known as Passchendale and Cambrai, took a similar amount of time and cost 400,000. By comparison the German spring offensive of 1918 lasted just over two months and resulted in 327,000 British losses. The "hundred days" of predominantly open warfare at the end of the war cost 364,000.

In short a month of "open warfare" was likely to cost 30 to 60% more men than a month during the set piece attacks. When the comparison is made with a period of fairly static "trench warfare" the contrast is even more marked: in the "quiet" period of four months from the beginning of December 1916 to the end of March 1917 the total losses were 77,000. The message is stark; "trench warfare" was comparatively cheap in pain and lives with total losses per day averaging perhaps 650 along the entire Western Front. Prolonged set piece attacks had a hefty "attrition" rate of about 2800; real "open warfare" was a totally unsustainable proposition, an Armageddon that gobbled about 4,200 in every 24-hour period.

At least for the enemy open warfare was not the promised land which it had once appeared to be, and it is questionable just how much the "triumph" of the "hundred days" at the end of the war was a product of the scientific application of new tactics, and how much a result of war weariness and demographic accident. Haig the high priest of the "wearing out battle," which we now call "attrition," claimed that the events of August, September and October 1918 were the logical conclusion of a hard fought but well thought out campaign. Yet numbers and logistics had their role to play. By the spring of 1918 the Germans were losing more men than they could recruit; and it might therefore appear likely, given the effects of the naval blockade, the slow but steady build up of American troops, and the allied tank superiority, that a breakthrough of meaningful proportions was becoming more rather than less likely.

After November 1918 the reverberations of the trench war would echo far into the future. A whole new generation of British Generals and politicians grew into post with a total aversion to the idea of a return to the slaughter of the trenches. Appeasement and unwillingness to go to war with Germany were the natural, if shortsighted, result. On the enemy side the whole political atmosphere would be tainted by the bitterness of defeat, outrage at the Treaty of Versailles, obsession with lost territory, the comradeship of veterans, and the ideals of the "Stormtrooper."

The experiences of the trenches also had a direct bearing on future military planning. The Germans had been on the receiving end of massed tanks, and gained a healthy respect for them. Though banned from tanks after the war, clandestine development of a German armoured force would become a high priority. Exercises with trucks began as early as 1921, and were followed by secret developmental work within the Soviet Union, beyond the eyes of Allied inspectors. By 1931 General Lutz was in post as Inspector of Motorised Forces, and a tank force command, or *Kommando der Panzertruppen* was set up in 1934. Many of the new ideas would be put

Right: **The lengthening war. The importance of the rum issue and the brazier to trench morale.**

Opposite page, left: **New heraldry of war. With British troops now clad in khaki Service Dress and scattered about the battlefield, battle patches or battalion identifying marks were now vital.**

Opposite page, right: **Formation signs and battalion badges of the British Empire: New Zealand, India, Australia, and Canada.**

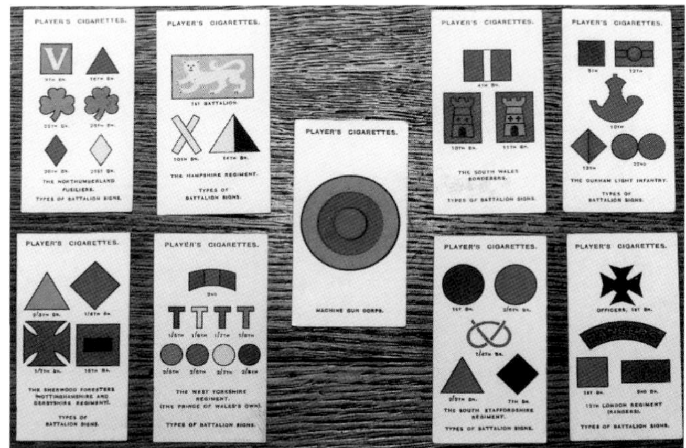

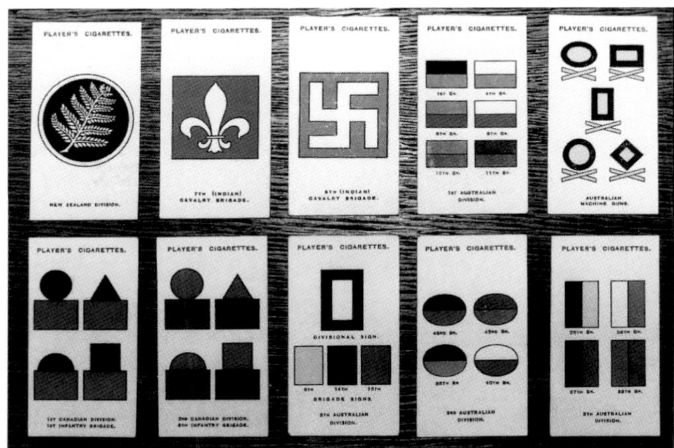

to the test in the Spanish Civil War, when the German Armoured Instruction Regiment of the Condor Legion participated on behalf of the Nationalists.

Just how much all this was influenced by the Great War, Allied armoured offensives, and British enthusiasts like J.F.C. Fuller, was eloquently explained in Heinz Guderian's *Achtung - Panzer !* of 1937,

"*It was now possible to move armoured vehicles and their crews unscathed by small arms fire until they closed with the enemy, and were able to bring them under direct fire and wipe them out. Motorised armoured vehicles also had the crushing capacity to cross and destroy the dreaded belts of barbed wire, and the obstacle-crossing capability to overcome obstacles and other obstructions. In late 1917 and in 1918 the true striking power of the Allied armies was therefore inherent in the tanks.*"

Ultimately the tactics of the Panzer division were those of the Mark V and Whippet turned Frankenstein's monster. Rather than have tanks "in penny packets" going short distances, the Germans now strove to complement the power of armour by the introduction of motorised supports which could keep pace with a concentrated body of tanks. In many instances, as in Poland and France, it would be the rest of the army and air force which attempted to coordinate with their actions. This "Blitzkrieg" prescription which would still have relevance in the Gulf War of 1991, when massed Allied armour, with the right artillery and air support, overcame Iraqi entrenchments with relative ease.

Nevertheless Germany still had resort to fixed defences. Construction of the "West Wall" in 1938 would see the pouring of six million tons of concrete. In the next two years 620,000 German workers were awarded the West Wall medal on a yellowish brown and white ribbon for their efforts. This work was doubtless a factor in deciding the Allies against attack in 1939, though how much was real and how much propaganda was not then put to the test. After the successful campaigns in the west the German Reich's new sea frontier received even more serious treatment in the form of the "Atlantic Wall." So was created what Hitler would call "Fortress Europe." Though patchy the Atlantic Wall certainly turned Allied invasion planners away from the Calais area to Normandy.

If the Germans had concentrated on developing what they perceived as the greatest enemy battlefield strength after 1918, then the same was true of the Allies. For the French this shibboleth was undoubtedly the power of trenches and concrete: and the Maginot line was essentially the Hindenburg line writ large. As early as June 1919 preliminary studies were commenced for a defensive system on the French Eastern frontier. Actual construction began in 1928, and was still ongoing after the outbreak of war in 1939. However, as we now know, this Leviathan defensive line was not equally strong along its course: cost and difficulties of construction forced the French planners to put most effort where the biggest threat was perceived to be.

In the south the Italian and Swiss borders were thought relatively safe, and so received little attention. North of the Swiss border, as far as Lauterbourg, the Rhine formed a significant obstacle and for the most part required only light fortification. The section from Lauterbourg to the Belgian border was perceived as the most problematic, for here an enemy might gain access to the important Lorraine industrial area. So it was here that the main effort on the Maginot line was concentrated. Parts of the works would be sunk 90 metres underground with layers of accommodation, stores, and gun rooms one above another. Entrances were well back behind the works themselves, and underground railways provided internal transport.

Belgium and France had a binding agreement to assist each other in the event of war with Germany, and the plan here was that French troops would enter Belgium and assist her to defend her Eastern border. It was therefore superfluous to plan significant static defences adjoining Belgium. Only in 1936 did Belgium decide to adopt a neutral status, and by this time the bulk of French finances had been committed elsewhere. Attempts would be made to improve this zone between 1936 and 1940, but nothing like the defences further south would be achieved.

Whilst the British did not carry the fetish of fixed defence quite so far, many still believed that the next war would be a "trench war" like the last. The result was considerable interest in pill boxes, and experiments with monster trench digging machines which were continued as late as the summer of 1941. Like the tank itself the development of the trench digger was pushed forward by Winston Churchill, who had indeed first suggested such a scheme during the Great War. Again like the tank, initial work was undertaken by the Navy, and was handed over to a Lincoln based contractor to develop and manufacture. As the "Naval Land Equipments" department was the driving force, the acronym "N.L.E." was applied to the trench diggers, which soon became the nickname 'Nellie'.

Originally it had been intended that 60 or more trench diggers should be made, with a maximum production of 20 units per week. The final design was for vehicles 77 feet in length, with two 600 horsepower engines, weighing about 130 tons. These giants would burrow forward at half a mile an hour cutting a trench five feet deep and seven feet six inches wide across the battle front, providing instant static cover. A pilot machine was actually produced and tested at Clumber Park Nottinghamshire, but the idea was overtaken by events. By 1942 work had ceased, and the project was finally wound up in 1943.

Perhaps more importantly the British strategic conduct of World War II was pervaded by a strong determination to avoid a repetition of the trench warfare, which had given war itself such a bad name in 1914-18. Defeat of the French in 1940 caused not to a return to the trenches, but retreat from Dunkirk. Stalin's demands for a second front before Churchill was ready, led not to a premature return to "the cockpit of Europe," but to sideshows and a major escalation of the bomber offensive. Yet when British and American troops did return to Normandy in 1944 the Bocage threatened to turn into full blown trench war. Indeed for a period of a few weeks that summer British infantry losses were much what they had been a generation earlier.

On the Eastern Front the huge distances, and the vast tank battle arenas of the Steppes made it unlikely that trench warfare would ever take hold. Even so there were exceptions to the rule. In the horror of Stalingrad, that 'Verdun on the Volga' a few square miles of battered town took on many of the attributes of the Western Front a generation before. Engineers with charges and flamethrowers became again more valuable than armour, and sniping developed into an art form and propaganda opportunity.

Despite the fame of Stalingrad the siege of Leningrad was of much longer duration, lasting for 900 days until finally broken by the Soviet offensive of January 1944. Though many escaped over the 'ice road' the siege was made all the worse by the large numbers of civilians who remained trapped. Exact numbers of fatalities are unknown, but the Soviet "Funeral Trust" alone buried 460, 000 civilian bodies in the year from November 1941 to November 1942. When the remaining time, and Soviet and German military dead are taken into account it is generally thought that the total death toll exceeded a million.

Though the Pacific battles were generally of much more limited duration, the Far East furnished some particularly nasty instances of trench warfare. The island fights of Iwo Jima and Okinawa for example were characterised by emplaced Japanese defenders who clung tenaciously to log bunkers, trenches, and caves. The Japanese had time to prepare, and made ample use of camouflaged snipers, emplaced howitzers, and mortars, but it was the Americans who had the technological advantage. The tanks, ground attack aircraft, grenades and flamethrowers all reduced American losses considerably: but there were instances where the enemy fought to the last man. In some particularly dif-

Top: **Multinational Allied prisoners, Hamm, July 1917. British, French, and Belgians are all represented in this one small group.**

Above: **German prisoners with British guards, High Lane Farm, near Sheffield.**

Left: **A German storm troop officer with a "concentrated charge," stick grenades, carbine, and 1916 model trench loop. (TRH Pictures)**

Opposite page, above: **Canadian wounded making a recovery, Sunningdale, Berkshire.**

Opposite page, below: **Signallers of the 3d Reserve Regiment of Hussars with flags and heliographs 1917. The traditional methods of communication were found woefully inadequate in trench warfare.**

ficult locations the Americans actually pumped underground tunnels with petrol and cremated the Japanese in their lairs.

In the new nuclear age a recurrence of trench warfare may have seemed unlikely, but peace in 1945 by no means put an end to position warfare. Being 'dug in' at the right moment continued to have as much relevance to the infantryman as it had done on the Western Front. Within five years United Nations forces would be facing North Korean and Chinese attack in Korea, and at times the conflict descended into trench warfare. This was made more likely where the hilly terrain precluded free use of armour, or when hand-held anti-tank weapons made tanks wary. "Human wave" attacks were generally no more successful than they had been in 1914, but the Communists also made skilled use of machine guns which were pushed ever closer to UN positions. As was predicted sub machine guns proved valuable in close conditions: but less expected was the unorthodox American use of bazookas as man portable artillery for anti personnel and bunker smashing work.

France would face its own trench warfare battles amongst the bunkers of Dien Bien Phu in Indo China in 1954. Having invested the area, the Viet Minh forces of General Vo Nguyen Giap took strong point "Beatrice' on 13 March following a heavy bombardment, and soon succeeded in closing the airstrip. At the end of the month the Communists launched a general assault. The French belatedly realised that a defensive position occupying a valley bottom would be extremely difficult to hold. Before long movement above ground became suicidal as the French paratroops and their local allies were reduced to occupying trenches and drainage ditches.

By the beginning of April the Viet Minh were lapping around the eastern strong points of 'Dominique' and 'Elaine'. Though Giap's men were experts at concealment and often remarkably courageous, their assault technique often lapsed into mass attacks which would not have been out of place on the Somme. Heavy bombardment at dusk would be followed by waves of troops, which were sometimes successful despite huge losses. Like the Germans in 1916 the French would launch immediate local counter attacks to try and regain ground: but unlike the Western Front Dien Bien Phu offered no obvious place to fall back. Parachute supply and reinforcement became a matter of life and death.

Failing to wrest control by means of mass attack the Viet Minh now determined to take the position by means of sapping forward, yard by yard, under the monsoon rains. The answer was raids, and the parachuting in of volunteers to the French cause including African and Arab soldiers. Nevertheless the perimeter was shrinking with the loss of the 'Huguette' and 'Elaine' positions. The French considered a final attempt at break out, but after 56 days General de Castries was forced to announce a cease fire. As one account put it, Dien Bien Phu had been "Hell in a very small place." The Viet Minh are believed to have suffered upwards of 20,000 casualties whilst the French

lost about half that number, plus another 6,000 taken prisoner. Only about 80 managed to melt through enemy lines to escape.

Though the Vietnam war is often thought of as either a jungle war, or a battle of high technology bombing against primitive courage and cunning, the conflict would offer several examples of trench warfare. At Nam Dong on 6 July 1964 a South Vietnamese armed camp was attacked resulting in a successful defence and the award of the first Congressional Medal of honour of the war to Captain R. Donlon. At A Shau in 1966 a Special Forces camp occupied by less than 400 men was set upon by 3,000 North Vietnamese regulars. After a 30 hour fight about half the defenders were rescued by helicopter. One American veteran remembered Khe Sanh as nothing but a "network of rat holes."

Often those resorting to tunnels and bunkers were the Viet Cong and North Vietnamese. In the face of heavy bombing, napalm, and American patrols, going into and under the ground was a natural reaction. Headquarters, operating theatres, and supply dumps were all built under the earth, sometimes protected by long winding tunnels which had to be crawled, multiple concealed entrances, and trap doors. In more than one location the South Vietnamese and the Americans occupied the surface whilst the enemy lived beneath their

Opposite page, above: **Souvenirs of the trenches. A No 9 Mark II folding box periscope, steel helmet, memorial plaque, Service and Victory medals, 1907 Pattern bayonet, and Princess Mary's 1914 Christmas fund box with tobacco.**

Opposite page, below: **Victory. Men of the Lincolnshire Regiment march through Cologne.**

Above: **Many Tommies never went home. A dead soldier of the North Staffordshire Regiment. In all 908,371 British and Empire soldiers were killed. For the Germans, who had run out of leather, removing boots from enemy dead became official policy.**

Below: **Lewis light machine gunners of the King's Own (Royal Lancaster) Regiment. Infantry tactics of small units based around light machine guns would be the new orthodoxy. Broadly similar ideas have survived to the present day.**

Above: **Postwar photograph of Field Marshal Haig. He was Commander in Chief of Home Forces in 1919–20, and involved in the Haig or "Poppy" Fund charity, as it became popularly known, until his death in 1928.**

Above right: **The legacy. Old soldiers 70 years after the Battle of the Somme, the cenotaph, St Helens, 1986. Left, Harry Holcroft DCM, 5th Battalion, South Lancashire Regiment; right, Dick Hesketh, Royal Artillery.**

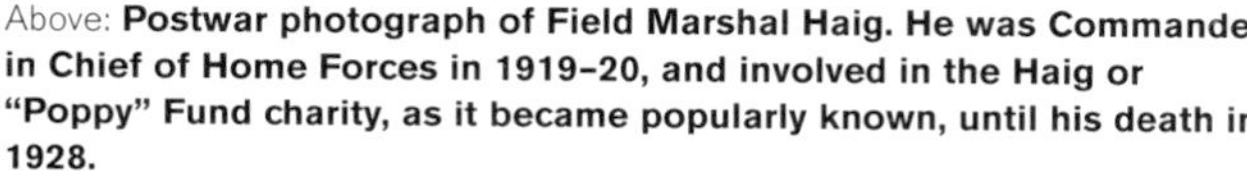

Right: **World War II and British troops are again dug in, this time on the Elbe. Occasionally, as in Normandy, or at the siege of Leningrad, full scale trench warfare would again dominate the fighting.**

feet. Units of "Tunnel Rats" would be formed who would attempt to prize out the Communists with grenades, fire, explosive charges, dogs, and revolvers.

Though full blown trench warfare in which emplaced positions dominate a whole campaign is now relatively rare, even the last twenty five years have seen occasions when stalemates have led to the revival of trench living. Perhaps the most celebrated example was the first Gulf War which began with Saddam Hussein's invasion of Iran in September 1980. Early Iraqi success stalled, and campaigning bogged down into attritional struggle. In July 1982 Iran launched her own Operation Ramadan when wave after wave of Revolutionary Guards threw themselves upon the enemy. The result was one of the biggest battles since World War II: but it ground to a halt without decisive result.

Iranian offensives were renewed in early 1983, and 1984, but the Iraqis, faced with a numerically superior enemy, resorted to familiar tactics. A complex web of defences 20 kilometres thick was developed, and a man made moat 24 kilometres long was flooded with water to defend Basra. Where there was little cover berms were built up, or trenches dug down. Where there was no infrastructure tar was poured directly onto the sand to create instant roads. Earth moving equipment became so important that Iraqi Third Corps mounted patrols specifically to destroy Iranian bulldozers.

Predictably the new offensives led to large casualties, but no end to the war. After four years of fighting Iran and Iraq had captured about 300 square miles of each others territory, for the loss of about 200,000, and 60,000 troops, respectively. As Arab commentator Mohammed Heikal explained;

'Mile after mile of mud, trenches, corpses, shell-holes and wrecked vehicles and artillery filled the landscapes of western Iran. Not since the battlefields of the Somme had there been so much carnage, sacrifice and courage on this scale. Fired with visions of martyrdom, a generation of Iranian youth hurled itself at Iraqi tanks and perished . . . Iraqi gunners worked ceaselessly behind earthworks thrown up by bulldozers, battering a human wall which renewed itself as soon as it was destroyed.'

In the second Gulf conflict the Iraqis tried to replicate the dense fortifications that had been so effective against the Iranians. On the Saudi border engineers planted thousands of mines, built berms and anti tank ditches, dug fire trenches, put up barbed wire, and set up heavy weapons to fire onto and over the obstacle zones. Armies totalling about 400,000 men backed the physical defences. Under water cables, mines, and wire decorated the Kuwaiti beaches.

This time however they would face a large well armed coalition, with overpowering air superiority,

special forces, and the latest tanks. In such a situation set piece defences could become as much a liability as a strength. In total 88,000 tons of munitions were dropped on the Iraqis. The hundred thousand sorties launched by Allied aircraft may not have inflicted massive casualties, nor destroyed more than a proportion of the field works; but they did cut communications and supplies, and gain total command of the air. Allied Multiple Launch Rocket systems proved more flexible than ordinary artillery as preliminary bombardment. Trench warfare would have no chance to take hold. Moreover, the Allied approach was not a frontal assault on Kuwait, but a hook through Iraq. This, as one commentator observed, hammered the enemy 'against their own defences', taking many positions from the side. As Lieutenant Alastair Stobie of the Royal Scots put it:

"*Going through was a bit of an anti climax really; you looked out and you could see what was going on around you. The Americans had literally bulldozed it flat; there weren't the trenches; the vehicles that were there were three-quarters buried and when you saw the top half of a T55 that was just peeled apart, destroyed, it was really good to know that someone previous to you had the firepower to do it.*"

Above: **German troops dig in, Arnhem, September 1944.**

Above right: **German troops digging in, 1940. The entrenching tool remains the soldier's friend even in mobile wars.**

Right: **The trench continues to be an important part of war. US and South Vietnamese troops, 1969. The weapons to hand are the 1918-vintage Browning Automatic Rifle and the M14.**

We should not however be sanguine, nor complacent. Short periods of trench fighting have occurred in both the former Soviet Union, and the former Yugoslavia. In 2001 in Afghanistan Taliban fighters occupied trench lines north of Kabul, and held off conventional ground attacks for a while. Thankfully precision bombing, politics, and logistics prevented this becoming a long term trench war. Yet war and holes in the ground have been intractable bed fellows for most of human history, and may continue to be so, despite international endeavour. On balance it seems far more likely that trench warfare will one day reoccur: for it lurks, quite literally, just below the surface of our common experience.

INDEX